Episodes From the Early
History of Astronomy

Springer
*New York
Berlin
Heidelberg
Barcelona
Hong Kong
London
Milan
Paris
Singapore
Tokyo*

Asger Aaboe

Episodes From the Early History of Astronomy

With 50 Figures

Springer

Asger Aaboe
Emeritus Professor of Mathematics, History of Science, and Near Eastern
   Languages and Civilizations
Yale University
New Haven, CT 06520
USA
asger.aaboe@yale.edu

Cover illustration: The frontispiece to Giambattista Riccioli, *Almagestum Novum* ...,
Bononia (Bologna), 1651. (Page 110)

Library of Congress Cataloging-in-Publication Data
Aaboe, Asger.
   Episodes from the early history of astronomy / Asger Aaboe.
       p. cm.
   Includes bibliographical references.
   ISBN 0-387-95136-9 (softcover : alk. paper)
   1. Astronomy—History. I. Title.
   QB15.A15   2001
   520'.9—dc21   00-061919

Printed on acid-free paper.

© 2001 Springer-Verlag New York, Inc.
All rights reserved. This work may not be translated or copied in whole or in part without the written permission of the publisher (Springer-Verlag New York, Inc., 175 Fifth Avenue, New York, NY 10010, USA), except for brief excerpts in connection with reviews or scholarly analysis. Use in connection with any form of information storage and retrieval, electronic adaptation, computer software, or by similar or dissimilar methodology now known or hereafter developed is forbidden.
The use of general descriptive names, trade names, trademarks, etc., in this publication, even if the former are not especially identified, is not to be taken as a sign that such names, as understood by the Trade Marks and Merchandise Marks Act, may accordingly be used freely by anyone.

Production managed by Allan Abrams; manufacturing supervised by Joe Quatela.
Typeset by Asco Typesetters, Hong Kong.
Printed and bound by Edwards Brothers, Inc., Ann Arbor, MI.
Printed in the United States of America.

9 8 7 6 5 4 3 2 1

ISBN 0-387-95136-9   SPIN 10780466

Springer-Verlag   New York Berlin Heidelberg
A member of BertelsmannSpringer Science+Business Media GmbH

For my grandsons,
Samuel and Tyler Alexander

Preface

More years ago than I can easily count, I published a small book entitled *Episodes from the Early History of Mathematics* (NML Vol. 13). Here I discussed in some detail a selection of subjects from Babylonian and Greek mathematics that could be fully mastered by someone with a background in high school mathematics.

My own work, particularly since then, however, has largely been concerned with ancient mathematical astronomy, especially Babylonian arithmetical lunar and planetary theories. I had the great good fortune in 1963 to get access to a large collection of unstudied, relevant clay tablets in the British Museum, so quite naturally I began thinking about writing an astronomical companion to the mathematical volume. I was well aware, though, that it would be quite a different sort of enterprise.

When I wrote the former volume, I could take for granted that my readers would have some familiarity with the elements of the subject whose early history was my concern, for nearly everyone has seen some basic arithmetic, algebra, and geometry.

It is, however, far otherwise for astronomy. If students take any astronomy at all, they may learn of the evolution of stars, and even of things that were certainly deemed unknowable in my youth, such as what Mars's surface and the far side of the moon look like in detail, but they remain in most cases woefully ignorant of what you can expect to see when you look at the sky with the naked eye, intelligently and with curiosity. I write this from experience. I first became aware of the state of astronomical enlightenment when, many years ago, out of curiosity I took a vote, yes or no, on a few questions from elementary spherical astronomy in a mathematics class of students selected for their excellence.

I first asked if the sun rises and sets. After the students' Copernican scruples were stilled, and they were sure I knew they knew that it was really the earth's rotation that made it so appear, they voted yes, the sun rises and sets. Next I asked the same question about the moon and, after some mutterings about phases, they voted yes again, as they did for the planets, for, as they reasoned, "planet" means "wandering star." But for the fixed stars their answer was a quick and unanimous no, for they are fixed.* My final question was whether one can see the moon in the daytime; again they voted no unanimously. This was disturbing, for the moon obliged by being plainly visible, and in broad daylight, through the lecture room's window, which I duly pointed out to them to their astonishment.

It was thus clear that an introduction to naked-eye astronomy would be necessary in a book on early astronomy; I have given one here in Chapter 0. I tried to keep it purely descriptive, uninfluenced by modern knowledge. A case in point is the retrogradation of planets. This phenomenon—that the planets, in their slow eastward motion among the fixed stars, come to a halt, reverse direction, and come to a halt again before continuing their eastward travel, is usually introduced in terms of the Copernican, heliocentric system. The argument involves the changing directions of lines of sight from a moving earth to a moving planet, and it is difficult, as I know only too well, to make students visualize, in this fashion, the phenomenon this explanation is supposed to explain. This is not strange, for planetary retrogression depends on the observer's being on the earth and so is best and easiest accounted for in a geocentric system.

In the astronomical introduction I simply describe the way the sun, moon, stars, and planets appear to behave to anyone who has the time, patience, will, and wit to observe and remember. Later in the book I discuss the various mathematical models, arithmetical and geometrical, that were devised in antiquity to account for the observed behavior of these bodies, particularly the planets.

---

*The correct answer is that some rise and set, while others never set, and the rest never rise. The two exceptions are at the equator, where all stars rise and set, and at the poles, where a star is either always above or always below the horizon. Thus, the students would have been nearly right if they had lived at either pole, but they did not.

I need not, however, introduce the mathematical techniques ancient astronomers used—Babylonian arithmetic and Greek geometry and trigonometry—for I already treated them in my previous little book.

In the following I concentrate on planetary theory and try to avoid the moon as much as possible. Here I am reminded of a story about Ernest Brown (1866–1938), professor of mathematics at Yale University. Brown devoted his life to the study of the moon's motion, and he published his lunar tables in three folio volumes in 1919. Toward the end of his life he was inanely asked what he could say about lunar theory. His answer was heartfelt, "It is very difficult." And so it has always been, ever since its elegant beginning in Mesopotamia some 2500 years ago, and I could find no place for it within the limits I had set for this little book.

Nor shall I consider early attempts at accounting for the planets' motion in latitude. These motions were referred to the mean sun, so their descriptions remained unduly complicated until Kepler saw that the planes of the planetary orbits pass through the true sun (it is in this connection he wrote that Copernicus was unaware of his own riches).

In Chapter 1 I introduce Babylonian arithmetical astronomy, the earliest, and highly successful, attempt at giving a quantitative account of a well-defined class of natural phenomena. I also mention the preserved observational records and hint at how the theories could have been derived from such material. [The elegant English mathematician G. H. Hardy (1877–1947) was particularly devoted to the theory of numbers—the domain of mathematics dealing with the properties of whole numbers—because of its purity in the sense that it found no application outside mathematics. He would have been greatly amused to learn that it was precisely number theory that, in the hands of the Babylonian astronomers, became the first branch of mathematics to be used to make a natural science exact.]

When I turn to Greek geometrical models for planetary behavior in Chapter 2, I can no longer afford to cite observations or to deal with how the quantitative models' parameters were derived from them. It is not that I deem these matters unimportant—quite the contrary—but such discussions would stray too far from my main purpose, which is to describe the various geometrical models for the planets and show how they work. Further, I am particularly interested both in demonstrating that epicyclic planetary models are not just ad-hoc devices for mimicking how a planet seems to behave, but are good descriptions of how a planet in fact moves relative to the earth, and in identifying their various components with their counterparts in the solar system. Indeed, a well-read medieval astronomer who considered a sun-centered system of planetary orbits could immediately derive its dimensions in astronomical units (one astronomical unit equals the mean distance from the earth to the sun) from the parameters of Ptolemaic epicyclic models, as we know Copernicus did (a note in his hand of this simple calculation is preserved in Uppsala University's library).

A crucial point in the demonstration is the transformation from a heliocentric to a geocentric coordinate system. This ought to be simple enough, but I have found it extraordinarily difficult to make people visualize the same phenomenon in the two systems. One sticking point is to make them realize that even the nearest fixed star is so far away that the directions to it from the sun and from the earth are the same (except for the annual parallax, which is so small that it was not observed until the 19th century).

The fixed-star sphere therefore looks the same whether viewed from the earth or from the sun. For an observer on the earth, the sun will seem to travel in a near-circular orbit, one revolution a year relative to the background of the fixed stars. The planets, in turn, will revolve in their orbits around the moving sun. This arrangement, the "Tychonic system" as Tycho Brahe himself modestly called it, is the one earthbound astronomers observe in and the one the Naval Observatory uses for compiling the *Nautical Almanac* (mariners, after all, are on the watery surface of the earth and do not care about what things look like from the sun).

The Copernican and Tychonic arrangements are geometrically equivalent, for either implies the other and yields the same directions and distances from one body to another. However, if you insist that the origin of your coordinate system—the earth for Ptolemy and Brahe, the mean sun for Copernicus (he has the true sun itself travel in a small circle around the mean sun)—is "at rest in the center of the universe," you do not, of course, have dynamical equivalence between the two systems, and the Copernican arrangement in Kepler's version, focused on the true sun, became the basis of Newton's mechanical treatment of the solar system.

In the *Almagest* Ptolemy does not give his planetary models absolute size: He measures a model's dimensions in units, each of which is one-sixtieth of the deferent's radius, because he only wants them to yield directions to the planets. So far they are perfectly compatible with the Tychonic system and, in fact, if we scaled the Ptolemaic models properly, they would also give us the distances to the planets correctly.

However, in his *Planetary Hypotheses* Ptolemy constructs his cosmological scheme, the Ptolemaic system, of snugly nested spherical shells, all centered on the earth, each containing the model of one planet, and with no wasted space. Here he commits himself to the dimensions of the structure in terrestrial units, and now agreement with the Tychonic system is no longer possible. I discuss these things in Chapter 3.

Though I originally intended to treat only ancient topics, I could not help including in Chapter 2 a few remarks on later modifications of Ptolemy's models at the hands of medieval Islamic scholars. These revisions were mostly of three kinds: improvements of parameters; much-needed corrections of serious flaws in his lunar theory; and attempts at replacing his philosophically objectionable equant with combinations of philosophically correct uniform circular motions that would work almost as well.

I end Chapter 2 with a few remarks about Copernicus and Brahe. Of Brahe's many achievements I only mention the Tychonic system, for its arrangement is, as said, compatible with the *Almagest*'s planetary models.

In my comments on Copernicus's work, I concentrate on just two aspects. First, I try to make clear precisely which problem was resolved by the heliocentric hypothesis. The general literature is often vague on this point, suggesting, for example, a desire for higher accuracy, or a dislike of epicycles, as a motivation for the new system. Both suggestions are wrong. In fact, the motivation lay in a desire to get rid of the awkward questions raised by the sun's curious role in the Ptolemaic planetary models: For an inner planet the deferent's radius to the epicycle's center always points toward the mean sun, while for the outer planets the radii from the epicycles' centers to the planets are all parallel and point in the same direction as that from the earth to the mean sun. This strange role was difficult to explain, but it becomes an immediate consequence of placing the mean sun in the middle of the planets' paths, as we shall see. (The Tychonic arrangement has the same virtue.)

Second, I discuss the fine-structure of Copernicus's planetary models. His ideal is clearly to have each planet move uniformly in a circular path—not, alas, around its center, but around an equant point in analogy to Ptolemy's deferent. However, he has committed himself to the exclusive use of uniform circular motions and manages, by superposition of several of these, to make his planets move uniformly around his ideal equant points—exactly, not approximately—while their resulting paths are nearly, but not quite, circular. I shall point out that these arrangements are precisely what we have found in the works of some of his Islamic predecessors.

In Chapter 3 I present, in some detail, Ptolemy's cosmology, the first cosmological scheme to include quantitative models in an integral way. It was long called the Ptolemaic system, even though Ptolemy's authorship of it was established only a few decades ago, when my then-colleague at Yale, Bernard Goldstein, found Ptolemy's own description of it in an Arabic translation of a lost part of the Greek original of his *Planetary Hypotheses*. Though the Ptolemaic system is now most often talked about in slighting terms, it is a logically pleasing structure that was, after all, the basis for how educated people thought about the universe for nearly a millennium and a half.

In the final chapter I show, from a more modern point of view, why Ptolemy's equant is so efficient. I take this opportunity to sketch how Kepler proceeded in order to find the longitude of a planet that moves according to his laws, and then I go on to analyze how a planet seems to behave when observed from the empty focus of its elliptical orbit, the one not occupied by the sun. It turns out that its angular motion is uniform but for terms involving second and higher powers of the ellipse's eccentricity, so the empty focus plays the role of an equant point and, for small eccentricities, ellipses are nearly circular. The combination of two eccentric circular motions with equants yields a planetary model very close to Ptolemy's.

The result of this analysis is not to be found in the more recent general literature, but it is far from new. In fact, in his *Principia* Isaac Newton addressed the problem of a planet's angular motion around the empty focus of its orbit and reached nearly the same result as the one I derive in Chapter 4.

In the course of this derivation I trangressed the limits I originally imposed on myself, for I could not help including an integral or two. However, having set the rules myself, I felt free to break them, and a few integrals never hurt anyone.

At one time or another, I have dealt with most of the above, very idiosyncratic selection of topics in courses and seminars at Yale University. Additionally, in March of 1988, I was pleased to be invited to give an Honors course at the University of Pittsburgh; in my lectures there I also presented some of this material. I am grateful to the students in both places who, by their questions, made what I thought and wrote clearer.

Furthermore, I wish to thank my colleagues, Dr. John Britton and Professors Bernard Goldstein and Alexander Jones, for reading my manuscript and for helpful criticism and suggestions.

Finally, I must acknowledge my indebtedness and gratitude to Miss Izabela Żbikowska of the Polish Academy of Science and Yale University. Without her constant help during the last three years, this little book, begun so long ago, would still be unfinished.

Acknowledgments

I have chosen special figures to illustrate the varied nature of the sources for our knowledge of earlier astronomy: cuneiform clay tablets and a Greek papyrus, fragmentary, but original; a page of a Byzantine manuscript, many copies removed from the original work; a page from the manuscript copy of *De revolutionibus* ... in Copernicus's own hand, about as authentic as evidence can get; and the frontispiece of an elegantly printed astronomical treatise of the 17th century.

The photographs of the cuneiform texts (Figures 2 and 5 of Chapter 1) are published through the courtesy of the Trustees of the British Museum.

The photograph of the Greek papyrus (Figure 6 of Chapter 1) is reproduced with the kind permission of the owner, Professor Marvin L. Colker.

The Vatican Library provided me with a photograph of Ptolemy's *Kinglist* from Vat. gr. 1291 and gave me permission to publish it (Figure 14 of Chapter 2).

The photographic copy of the frontispiece of Riccioli's *Almagestum Novum* ... of 1651 (Figure 20 of Chapter 2) was made for me by members of the Institute for the History of the Exact Sciences at Aarhus University, Denmark, the home away from home for so many scholars, from a copy in their library.

I offer my thanks to these people and institutions.

Advice to the Reader

I have arranged the chapters in approximate chronological order, but each makes some sense even if read independently. Some readers might profitably begin, after Chapter 0, with Chapter 2 on Greek geometrical models, for they may find its methods less strange than those of the Babylonian arithmetical schemes.

New Haven, Connecticut   Asger Aaboe
January, 2001

Contents

Preface		vii
0	What Every Young Person Ought to Know About Naked-Eye Astronomy	1
1	Babylonian Arithmetical Astronomy	24
2	Greek Geometrical Planetary Models	66
3	Ptolemy's Cosmology	114
4	Kepler Motion Viewed from Either Focus	135
Selected Bibliography		171

0
What Every Young Person Ought to Know About Naked-Eye Astronomy

In order to provide a starting point for an understanding of ancient astronomical texts, I shall begin by presenting, in all brevity, the basic elements of naked-eye astronomy. I shall, of course, deal principally, but not entirely, with phenomena of interest to ancient astronomers. Among these are many phenomena, such as the first or last visibility of a planet or the moon, that the modern astronomer shuns since they take place near the horizon and further depend on imperfectly understood criteria. Thus, these phenomena are not commonly discussed in the modern astronomical literature and, more seriously, we lack modern standards with which we may measure the quality of the ancient results.

My presentation is entirely descriptive. I do not attempt to explain why planets become retrograde, or that celestial bodies really do not rise and set but that the earth's rotation makes it appear that they do. Whoever does not feel comfortable talking about the solar system without taking a detour via the sun is welcome to do so.

The Celestial Sphere, Fixed Stars, Daily Rotation

We have no simple means of judging distances to celestial bodies; we can only determine the directions toward them. To describe what we thus observe, we introduce the *celestial sphere* as a spherical surface with its center at the observer's eye and of unit radius—which unit is irrelevant, but it is often thought more comfortable to choose a very large one. A celestial object is then mapped at, or identified with, the point on this spherical surface at which the line of sight to the object pierces the celestial sphere. The study of the behavior of this map of celestial objects on the celestial sphere is called *spherical astronomy*.

If all celestial bodies visible to the naked eye are thus mapped on the celestial sphere, it becomes apparent that the vast majority of them remain in fixed patterns with respect to each other: They form recognizable constellations that, in turn, remain unchangeably distributed. These celestial bodies are called the *fixed stars*. To the naked-eye observer there remain seven exceptional objects: the Sun, the Moon, and the five bright planets: Mercury, Venus, Mars, Jupiter, and Saturn; being exceptional, they are, of course, of particular interest.

Before describing the celestial sphere and its motion, I shall briefly introduce, informally and without proofs, some basic terminology and a few results from spherical geometry (i.e., the geometry on the surface of a sphere).

A plane cuts a spherical surface, if at all, in a circle (or in one point if the plane is tangent to the sphere). If the plane happens to pass through the center of the sphere, the section is called a *great circle*; otherwise the section is called a *small circle*. Great circles are fundamental to spherical geometry; indeed, they play much the same role that straight lines do in plane geometry. Thus, through two points on a spherical surface that are not diametrically opposite, passes one and only one great circle, and the shortest distance between two such points measured on the sphere is along the great circle joining them. By the "distance between two points on the sphere," we mean the shortest distance, namely, the length of the shorter of the two great circle arcs that join them; this distance is usually given not in linear measure but in degrees (thus, the longer of the two great circle arcs joining two points will be 360° minus the distance between the points).

Associated with a great circle are two points called its *poles*; they are the end points of that particular diameter of the sphere which is perpendicular to the plane of the great circle (i.e., they are to the great circle what the North and South poles are to the equator). Conversely, to two diametrically opposite points on the sphere there corresponds one, and only one, great circle whose poles they are. The distance from a pole to any point on the corresponding great circle is 90°.

Two great circles always intersect in diametrically opposite points, that is, they always bisect each other. The angle between two great circles is the same as the distance (in degrees) between their poles.

These few remarks may suffice for our present purposes, and we can now return to the celestial sphere and its behavior.

The fixed stars permit us to get a hold on the celestial sphere—they provide us with a coordinate system, if you will—and enable us to perceive its motion. Ignoring for the moment various very slow changes, we will observe that the celestial sphere, with the fixed stars fixed upon it, revolves about a fixed axis at a fixed rate of very nearly 366 1/4 (*sic*) revolutions per year, relative to familiar fixed objects in our surroundings. This axis, being a diameter of the celestial sphere, pierces it at two diametrically opposite points called the celestial *north* and *south poles* (the north pole is now very near the North Star, or Stella Polaris). The great circle corresponding to these poles (i.e., the great circle that slides in itself during the *daily rotation*, as this motion is usually named) is called the *celestial equator* or, simply, the *equator*.

A horizontal plane through the observer intersects the celestial sphere in a great circle called the *horizon*, and a vertical line, also through the observer, pierces the sphere at two points, *zenith* above and *nadir* below, which are the poles (the term "pole" is used here as in spherical geometry) belonging to the horizon. Celestial objects below the horizon are invisible since the line of sight to them would pass through the body of the earth: The plane of the horizon is ideally tangent to the spherical earth.

The vertical plane through the observer and the north pole, which because it is vertical also contains the zenith, intersects the celestial sphere in a great circle called the *meridian* and meets the horizon in two points: the *north* and *south points*.

The vertical plane through the observer perpendicular to the plane of the meridian intersects the celestial sphere in a great circle called the *first vertical*, which meets the horizon in the *east* and *west points*. The equator passes through the east and west points.

For a given place of observation, the directions to the north, south, east, and west points as just defined, remain fixed in relation to characteristic features of the visible neighboring terrain. Furthermore, the elevation of the north pole does not change.* These two facts together imply that the axis through the north and south poles, about which the daily rotation takes

---

*Here there is no need for the cautionary remark about long-term changes.

place, remains fixed for an observer at a given locality in relation to his or her terrestrial surroundings. The elevation of the north pole above the horizon, in angular measure, is called the terrestrial latitude of the place of observation and is usually denoted by ϕ. For Babylon we have $\phi = 32\frac{1}{2}°$, very nearly.

In relation to a given horizon, the fixed stars are divided into three classes: those that are always above the horizon; those that are sometimes above and sometimes below the horizon; and those that are always below the horizon.

The stars in the first category are called *circumpolar*. They are all within a cap with the north pole as its center (for observers in the northern hemisphere of the earth) and a radius equal to the observer's terrestrial latitude ϕ. Thus, the farther north you live, the larger the region of the circumpolar stars.

A corresponding cap of equal size but centered on the south pole contains the fixed stars that are never above the horizon.

All the fixed stars in the belt between these two circumpolar caps will cross the horizon in the east and in the west and by the daily rotation will be carried in a path partly above and partly below the horizon. This daily path—a circle on the celestial sphere—is traversed by the fixed star in slightly less than 24 hours. A star on the equator is as long in time above as below the horizon, for its diurnal path is bisected by the horizon. Again for an observer in the northern hemisphere, if a star is north of the equator, it spends more time above the horizon than below—the difference is larger the nearer the star is to the circumpolar cap—and symmetrically for a star south of the equator.

Since a star is visible only at night and when above the horizon, it follows that circumpolar stars are visible every night of the year. Those circumpolar about the south pole are never visible. A star in between is visible more nights the farther north it is, for the interval of nighttime shifts throughout the year in relation to a star's horizon crossings, as we shall see.

The sizes of the circumpolar caps and of the horizon-crossing belt vary with terrestrial latitude, as mentioned. To illustrate this, let us consider two extremal situations.

First, for an observer on the earth's North Pole, where the terrestrial latitude ϕ is 90°, zenith and north pole coincide. The circumpolar caps meet at the horizon, and the horizon-crossing belt vanishes. All visible stars are circumpolar, their diurnal paths are parallel to the horizon, and no stars ever cross the horizon. One has a chance of seeing only the stars on the northern celestial hemisphere, whereas the rest are never above the horizon.

Second, for an observer on the terrestrial equator, it is the circumpolar caps that vanish while the horizon-crossing belt fills the entire celestial sphere. The north and south poles are in the horizon, and all stars spend equal time above and below the horizon. One has, theoretically, an equal chance of seeing all stars, though in practice those stars near the poles never get very far above the horizon.

The diurnal circle of a star in the horizon-crossing belt intersects the meridian in the points of *upper* and *lower culmination* at which the star is at its greatest elevation above, and lowest depression below, the horizon. For a circumpolar star these two points mark its greatest and smallest distance from the horizon, respectively.

Sun, Ecliptic, Seasons

For two reasons it is natural to begin an introduction to spherical astronomy, as I have, with the fixed stars. First, their behavior is simpler than that of the other celestial bodies: They remain fixed in relation to each other, and all join in the uniform diurnal rotation about a fixed axis. Second, they provide a convenient background against which the more complicated behavior of sun, moon, and planets can be perceived and described. It must be emphasized that once a celestial body has been placed among the fixed stars it will, of course, partake of the same daily rotation that they are subject to, in addition to any motion that it will have relative to them.

The first example of this approach is the case of the sun. Let us imagine, contrary to our everyday experience, that it were possible to see the sun and fixed stars at the same time. (This is now so for an observer outside the earth's atmosphere.) We would then note that the sun, day by day, moves eastward very slowly among the fixed stars, in the amount of about 1° per day. It returns exactly to its original place after the lapse of one year—that is, indeed, the definition of the year or, more precisely, the *sidereal* year—having traced out in that interval a path among the stars which is a great circle. Year after year the sun travels precisely the same great circle, which is called the *ecliptic*.

In the time it takes the sun to complete one revolution in the ecliptic, namely, in one year, the fixed stars revolve, as said, very nearly 366 1/4 times relative to, say, the meridian. This is the same as saying that a year has about 365 1/4 days, for the sun has in that period revolved the same number of times as the stars less the one revolution it itself has performed relative to them, but in the opposite sense of the daily rotation.

The ecliptic plays a fundamental role as reference circle in ancient astronomy. This is not surprising, for not only is it the path of the sun, but the moon and the planets are always within a belt extending at most 10° on either side of it.

The ecliptic has an inclination toward the equator of about $23\frac{1}{2}°$. The two diametrically opposite points of intersection between ecliptic and equator are called the *equinoxes*. When in its travel the sun happens to be in either one of these, its diurnal path is the equator itself which is bisected by the horizon, so day equals night in duration. When in its yearly motion the sun crosses the equator from the south to the north, it is *vernal* or *spring equinox*—this term is commonly applied both to the phenomenon and to the point on the celestial sphere—and the other crossing is called *autumnal* or *fall equinox*.

The point halfway between the equinoxes and at which the sun is farthest north of the equator is the *summer solstice*, and the diametrically opposite point is the *winter solstice*. When the sun is at these points, the duration of daylight is longest and shortest, respectively.

Let us, once more, consider the previous two extremal situations. First, for an observer on the North Pole of the earth, where zenith and celestial north pole coincide, as do horizon and celestial equator, the sun will appear in the horizon for the first time at "vernal equinox" and will not set until "autumnal equinox." From the vernal equinox the sun slowly gains elevation above the horizon until it reaches an altitude of $23\frac{1}{2}°$—the amount of inclination of the ecliptic against the equator—at summer solstice. In the course of 24 hours the sun will be seen above every point of the horizon, and the shadow cast by a vertical stick will revolve 360°. From summer solstice the sun slowly works its way back down to the horizon, and from fall equinox to spring equinox it will be invisible.

On the earth's equator, where the celestial equator and the first vertical coincide, day and night are always equal, and the "equinoxes" are marked by the sun passing through zenith at noon. At the solstices the sun crosses the meridian farthest north or south of zenith.

Two further special cases are of interest. One is to find the zones on the earth where the sun becomes circumpolar just once a year. For this to happen the circumpolar cap whose radius is the terrestrial latitude ϕ must reach the sun when farthest (i.e., $23\frac{1}{2}°$) from the equator. Thus, ϕ must be $90° - 23\frac{1}{2}° = 66\frac{1}{2}°$. At this northern latitude the sun will just reach, but not cross, the horizon at midnight on summer solstice, and symmetrically for the southern hemisphere. The two circles on earth of these latitudes are called the *polar circles*.

The other special case is to determine where on the earth the sun just reaches the zenith once a year at solstice. It is readily seen that this happens at terrestrial latitude $23\frac{1}{2}°$ north or south. The corresponding two circles are called the *tropics*, the northern *of Cancer*, the southern *of Capricorn*, for reasons that will become clear later.

Finally, I shall mention a variant way of characterizing terrestrial latitude, namely, by giving the ratio of longest to shortest daylight. This works, of course, only for places between, but not on, the polar circles. Since, for reasons of symmetry, the shortest day at a given locality equals the shortest night in length, this ratio tells us how the sun's diurnal circle at summer solstice is divided by the horizon; it is then a fairly simple matter to determine, if one wishes, the elevation of the pole (i.e., the place's terrestrial latitude) by means of spherical trigonometry.

On the equator of the earth this ratio is 1:1. The farther north a locality, namely, the higher the North Pole is elevated above the horizon, the larger this ratio becomes until it loses definition on the polar circle, where the longest day is 24 hours and the shortest 0 hours.

We find this practice in ancient Greece, and a ratio of longest to shortest daylight of 3:2 for Babylon is, as we shall see, a fundamental parameter in Babylonian astronomy. I must emphasize, however, that we have no evidence whatsoever that the Babylonians were aware that this ratio changes as one travels north or south.

Synodic Cycle of a Star Near the Ecliptic

As an introduction to phenomena of the kind dealt with in Babylonian astronomy, let us consider a fixed star on or near the ecliptic; we shall be concerned with when, where, and how long it is visible from a place of observation of reasonable latitude.

Since the star is close to the sun's yearly path, at a particular time of year the sun and star nearly coincide on the celestial sphere or, as we say, are in conjunction. The star will be invisible, for whenever it is above the horizon, so is the sun. Star and sun will rise simultaneously on that day.

When the star is about to rise the next morning, the sun will have moved about 1° farther along the ecliptic so that the star rises a little before the sun. The following morning the sun will have moved yet another degree away from the star, so the interval from starrise to sunrise has lengthened. Eventually a morning will come when the star rises so long before sunrise that the sky is sufficiently dark for the star to be visible as it crosses the horizon, if only for a short while until the dawn extinguishes it. This is the phenomenon of *first visibility* (we usually denote it by the letter Γ).

From the morning of first visibility, the star will rise earlier and earlier, and we can follow it farther and farther from the eastern horizon along its diurnal arc before dawn makes it vanish.

Half a year after conjunction the sun reaches a point on the ecliptic 180° from the star. Since ecliptic and horizon, being great circles, always intersect in diametrically opposite points, we now have the situation that when the star is in the horizon, so is the sun. Thus, the star rises at sunset and sets at sunrise; further, the star will be in upper culmination at midnight. We say that now the star is in *opposition* to the sun, and we use the letter Θ for this phenomenon. We are able to see the star all night long—less the intervals of dawn and dusk.

As the sun now progresses in its yearly motion, it approaches the star from the other side, from the west. After each sunset, when it gets dark enough to see it, the star will already be well past the eastern horizon, and it will at its appearance get closer and closer to the western horizon across which it sets.

There will now come an evening when the star appears only just before it sets, while on the next evening the sun will have gotten so close to it that the sky is not dark enough for the star to be seen before it sets. This is the phenomenon of disappearance of the fixed star, and we denote it by the letter Ω.

The star will now remain invisible while the sun catches up with it. Star and sun will, once again, be in conjunction, and the cycle is closed, having occupied just one year.

What I have described here is called the fixed star's *synodic cycle*, consisting of the *synodic phenomena* conjunction, first appearance, opposition, last appearance, and conjunction—once again, all phenomena that place the star in special relation to the sun.

The visibility phenomena, first and last appearance, or Γ and Ω, are determined by several factors, some of which are difficult to control. First they depend on the brightness of the star: The brighter the star is, the shorter the sun has to be removed from it for it to be visible near the horizon. Second, they depend on the inclination of the ecliptic against the horizon when the star is rising or setting: The smaller the inclination, the farther the sun must be from the star to ensure the darkness necessary for visibility. Furthermore, these phenomena depend on the acuity of the observer's eyesight and the quality of the atmosphere; in the former there is clearly a personal variation, and the latter is difficult, if not impossible, to ascertain for a place of observation in antiquity.

Since the synodic period of a fixed star is one year, any one of the synodic phenomena can be used as a seasonal indicator. This is, indeed, an ancient practice. It is well known that the first appearance of Sothis, our Sirius, was used by the ancient Egyptians as a herald of the rising of the Nile—two seasonal phenomena that happened to coincide—and one finds various rustic tasks tied to first or last appearances of certain fixed stars or constellations in primitive societies as we learn, for instance, from Hesiod's *Works and Days*.

Finally, a remark about the term "synodic"; *synodos* in Greek means "getting together" or "meeting" and in astronomy is used particularly for the coincidence in position of a celestial object with the sun, that is, for conjunction. It is now used for any phenomenon linking a star, a planet, or the moon to the sun in a certain fashion.

Synodic Cycle of an Outer Planet

The planets are conspicuous among celestial bodies for at least the following three reasons: They are all very bright; they do not twinkle like the fixed stars; and they move relative to the fixed stars though always staying close to the ecliptic. They are divided into two classes consisting respectively of those that can reach opposition to the sun and those that cannot.

The members of the first class—we call them *outer planets*—are Mars, Jupiter, and Saturn, as far as naked-eye astronomy is concerned. The second class consists of the two *inner planets*, Mercury and Venus. Characteristic of them is the limit to how far they can get from the Sun: Venus never gets farther from the Sun than about 40°, and Mercury's limits are even narrower—about 25°. We shall first be concerned with the outer planets.

The synodic behavior of Saturn has much in common with that of a fixed star near the ecliptic. Like it, Saturn makes its first and last appearance (also denoted Γ and Ω), and near the middle of its interval of visibility it is in opposition (Θ) to the sun. The difference is that Saturn has a motion of its own: The general trend of this motion is that from Γ to Γ (or from Ω to Ω), that is, during one synodic period, Saturn moves about 12° along the ecliptic in the same direction—eastward—as that of the Sun's proper motion. This means that the synodic period of Saturn is some 12 days longer than a year, for the Sun will have to travel 12° in excess of one complete revolution in the ecliptic to catch up with Saturn again, and the Sun travels almost 1° per day.

Within each synodic period Saturn's motion is, however, more complicated; after its first appearance Γ, Saturn first moves *directly*, as we call eastward motion along the ecliptic, then comes to a halt among the fixed stars—it is now at its *first stationary point* Φ—then reverses its motion, or becomes retrograde, and then comes to a halt once again at its *second stationary point* Ψ, where its motion finally becomes direct, once again. Opposition happens in the middle of the retrograde arc. The synodic phenomena of Saturn are then, in their proper order,

Γ: first visibility
Φ: first stationary point ⎫
Θ: opposition   ⎬ retrogradation
Ψ: second stationary point ⎭
Ω: disappearance

An actual run of Saturn for some three synodic periods is illustrated in Figure 1, where the dimensions perpendicular to the ecliptic are exaggerated four times for the sake of clarity. The dotted stretches from Ω to Γ are its arcs of invisibility in the middle of which it is in conjunction with the sun.

The other outer planets, Jupiter and Mars, behave qualitatively like Saturn. The character and sequence of their synodic

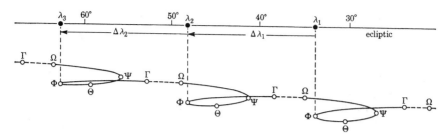

FIGURE 1.

phenomena are quite the same, but their *synodic arcs*—the progress from, say, Γ to Γ—are larger. Jupiter's synodic arc is some thirty-odd degrees, and that of Mars is even in excess of 360°. Thus, their synodic periods are longer than Saturn's, that of Mars even longer than two years. I shall later return to these matters and treat them from a more quantitative point of view.

Synodic Cycle of an Inner Planet

Since an inner planet is, as it were, tethered on a short leash to the sun, it will, in the long run, travel among the fixed stars as many times as the sun does or, on the average, once a year.

When an inner planet is, within its bounds, sufficiently west of the sun, it will be visible in the eastern sky for an interval before sunrise, and it is then called a *morning star*. When it is sufficiently east of the sun, it will be seen in the western sky for a while after sunset, and it is then called an *evening star*. Since Mercury is tied to the Sun within rather narrow limits, it is visible, if at all, only for relatively short intervals before sunrise or after sunset. The synodic phenomena of an inner planet are, in order,

Ξ: first appearance as an evening star; greatest eastern elongation from the sun
Ψ: stationary point in the west ⎫
Ω: disappearance as an evening star; inferior conjunction ⎬ retrogradation
Γ: first appearance as a morning star
Φ: stationary point in the east; ⎭
greatest western elongation from the sun
Σ: disappearance as a morning star; superior conjunction

The phenomena that have not received Greek letters are not considered in Babylonian astronomy. At inferior and superior conjunction, an inner planet is, of course, invisible so it is not odd that they are not among the Babylonian synodic phenomena (the distinction between inferior and superior conjunction reflects our modern knowledge that the inner planets circle about the sun). One may find it strange, however, that an inner planet at greatest elongation from the sun—the situation when it is visible the longest—is not given special consideration.

The Moon's Synodic Course

The moon moves directly, namely, eastward, among the fixed stars at a rate of about 13° per day in an orbit that is inclined approximately 5° toward the ecliptic; the moon will thus return very nearly to the same position among the fixed stars after some $27\frac{1}{3}$ days, an interval that is called one *sidereal month*.

To get some sense of the moon's swiftness, one may consider that the moon's apparent diameter is slightly more than $\frac{1}{2}°$, so the moon moves a distance of its own width in less than one hour. Thus, the moon's proper motion can be readily perceived in quite a short time if one refers it to neighboring fixed stars.

When we considered the synodic cycle of a fixed star or of an outer planet, the sun was the swifter body that ran away from, and caught up with, the other. With the moon the roles are reversed. If sun and moon are in conjunction near a fixed star at a certain moment, the moon will reach the star again after one sidereal month, or some $27\frac{1}{3}$ days, in the course of which the sun has moved ahead only some 27°. Thus, it will take the moon a little more than two extra days to reach the sun and, once more, be in conjunction. The interval in time from conjunction to conjunction is, then, somewhat over 29 days—on the average, 29.5309 days, to be more exact—and is called one *synodic month* or one *lunation*. In addition to the year and to the day, this is a most important time interval in ancient calendars, and fundamental in Babylonian astronomy.

When we now consider the sequence of synodic phenomena within each synodic cycle, we should recall that the moon's phase can be used as an additional indicator of its synodic state. Furthermore, it should be remembered that the moon is the only celestial body that may be seen with ease simultaneously with the sun, that is, in the daytime (one may actually at times follow the progress of Venus across the day sky, but it is difficult, and one has to know where to look).

At conjunction the moon is invisible, unless it happens to eclipse the sun; we shall return to the question of eclipses later. As the moon now moves away from the sun at the rate of some 12° per day—the difference between lunar and solar daily motion—it increases its chances for being visible in two ways: by getting farther from the sun, and by its lighted sickle growing in width. At the first or second sunset—in exceptional circumstances, perhaps the third—after conjunction, the thin crescent becomes visible in the west against the darkening evening sky before the moon sets following the sun. This is the *evening of first visibility* of the new moon and it marks the beginning of a new month in the Babylonian calendar. Determining in advance which evening the new moon will become visible is one of the chief goals of Babylonian lunar theory and, as we shall see, is a highly difficult task.

Some seven days after conjunction, the moon is 90° from the sun, and half its visible surface is lighted. We say that the moon is at *first quarter*. It will now rise near noon, when the sun culminates; at sunset the moon at first quarter will be near culmination, and it sets near midnight.

Half a synodic month, or 14 to 15 days, after conjunction, the moon reaches opposition to the sun. Its entire visible surface is lighted and we have *full moon*. The full moon rises at sunset and sets at sunrise. Eclipses of the moon occur at full moon. Both conjunctions and oppositions of the moon are called by the common term *syzygy* (strictly speaking, "syzygy" means precisely the same as conjunction, namely, the state of being yoked together).

Some 21 or 22 days after conjunction, the moon is at its *last quarter*, and a little before the next conjunction will be a morning when the waning sickle—the term "crescent" is still often used, though a misnomer—is seen to rise just before sunrise for the last time. This is the *morning of last visibility* of the moon, and it is followed by an interval during which the moon is too close to the sun, and its lighted part too slender, for it to be visible. When conjunction is reached in the middle of this interval of invisibility, the synodic cycle of the moon is closed.

To recapitulate, the moon's synodic phenomena are, in order,

conjunction
first visibility
first quarter
full moon or opposition
last quarter
last visibility
conjunction

As said, the term *synodic month* is applied to the time interval from one synodic phenomenon to the next synodic phenomenon of the same kind. But there is a certain difference to which I ought to draw attention. If the phenomenon is a syzygy—conjunction or opposition—a specific synodic month will be some fraction of a day in excess of 29 days; just how much will depend on various factors, as we shall see. But when the phenomenon is one that, like first visibility, can happen only at a certain time of day, here sunset, any synodic month must be a whole number of days—for first visibility either 29 or 30 days. In either case the *average* length of the synodic month will, of course, be the same, namely, 29.5309 days.

Lunar Orbit and Nodes

In the preceding discussion we ignored that the moon's orbit is inclined to the ecliptic, for the inclination is so small that it plays no significant role in a qualitative description of the moon's synodic phenomena. However, where the moon is in its inclined orbit can be decisive when one wishes to determine, for example, precisely on which evening the moon will first become visible, and for the prediction of eclipses it is absolutely essential to know how far away from the ecliptic the moon is.

The lunar orbit is, then, a great circle on the celestial sphere which is inclined to the ecliptic at an angle of ca. 5°. The two diametrically opposite points of intersection between the ecliptic and the moon's orbit are called the *lunar nodes*; the one where the moon crosses the ecliptic from the south to the north is called the *ascending node*, the other the *descending node*.

The first difficulty is that the nodes do not remain fixed among the fixed stars, but have an appreciable retrograde motion of not quite 2° per synodic month (this motion does not affect the orbit's inclination). Consequently, if the moon, a fixed star, and the ascending node happen to coincide on the celestial sphere at a certain moment, the moon will return to the ascending node almost 3 hours before it reaches its nearest approach to the fixed star, for the node has moved so as to meet it earlier. The period of the moon's return to a given node is called the *draconitic* (or draconitc, or nodical) *month* and it is a trifle more than $27\frac{1}{5}$ days in length.

Some 7 days after the moon has been at the ascending node, it will be as far north of the ecliptic as possible, namely 5°; after a further 7 days or, more precisely, after in all half a draconitic month, the moon will be at the descending node; after still another 7 days the moon will be as far south of the ecliptic as it can get; and after the lapse of a full draconitic month it will, once again, be at the ascending node.

Eclipses

A lunar eclipse happens when the earth intervenes between the sun and the moon and deprives the moon of the sun's light or, in other words, when the moon enters the earth's shadow. For this to take place two conditions must surely be satisfied: First, sun and moon must be seen in opposite directions by an observer on the earth, that is, the moon must be in opposition, or be full; second, since the line from the observer to the sun is always in the ecliptic, and since sun, earth (observer), and moon must be in a straight line, the moon must be at, or near, a node.

A lunar eclipse is, of course, a *real* eclipse in the sense that anyone who can but see the moon at the time—even an observer on the moon itself—will notice that the moon is deprived of the sun's light. For observers on the earth, this means simply that if only the eclipse occurs during their night, they should be able to see it.

A solar eclipse happens when the body of the moon intrudes between the observer and the sun. The character of a solar eclipse is thus quite different from that of a lunar eclipse. A solar eclipse is a more subjective phenomenon; for an observer somewhere in space it is signalled not, of course, by any change in the aspect of the sun, but by the appearance on the lighted surface of the earth of a black spot, the shadow cast by the moon. The conditions that ensure that a solar eclipse will occur at a given locality are, then, first the necessary ones that sun and moon are in conjunction, that the moon is sufficiently close to a node for its shadow to hit the earth, and that this happens while it is day at the place of observation. In addition, one must ascertain that the locality lies in the path of the moon's shadow. To do this, one must know, at least, the shape of the earth and the relative sizes and distances of sun, moon, and earth. If I may introduce a historical remark, Ptolemy's *Almagest* (ca. A.D. 150) is the earliest surviving work in which this problem is reasonably dealt with and on the basis of which one may predict a solar eclipse for a particular place with confidence. While the Babylonian texts yield solid predictions of lunar eclipses, they serve only to present necessary, but not sufficient, conditions for solar eclipses or, if you wish, to issue warnings, but not predictions, of solar eclipses.

The simple, necessary condition for solar as well as lunar eclipses is then that the moon at syzygy be near a node. A rough estimate of how often this happens is readily reached. Let us assume that a syzygy of a certain kind, say, a conjunction of sun and moon, takes place at a node, say, the ascending one. The moon will reach that node again after the lapse of a draconitic month, which, on the average, is 27.21 days. However, the moon will not catch up with the sun again until after a full synodic month, which, in the mean, is 29.53 days. At the next conjunction the moon will then be 2.32 days' travel past the ascending node. At the second conjunction, the moon will be twice that distance from the ascending node, and so on. At the sixth conjunction after the original one, the moon will be $6 \times 2.32 = 13.92$ days' travel from the ascending node; however, since half a draconitic month, or 13.61 days, brings the moon from the ascending to the descending node, the sixth conjunction will take place just past the descending node. Thus, six synodic months bring a conjunction—or an opposition—of sun and moon from one node to a little past the other, and eclipse possibilities will then happen at intervals of mostly six, but occasionally of only five, synodic months. An interval of only five months must intervene at times since the six-month interval is clearly a little too long and since, further, experience shows that two eclipses for a given locality never take place only one month apart. (Modern canons do show *solar* eclipses at an interval of only one month, but in general one of these will be visible only near the earth's North Pole and the other only near its South Pole.)

It is then proper to issue eclipse warnings at intervals of mostly six, but occasionally of only five, months; what one can guarantee is not, of course, that eclipses will happen at these syzygies, but rather that eclipses will not occur at any other. This holds for both solar and lunar eclipses.

Coordinate Systems and the Zodiac

So far I have tried to keep the discussion of spherical astronomy as qualitative as possible, but as the need arises to become more precise and quantitative, we must introduce coordinates.

The basic device for fixing a position on a spherical surface by a pair of numbers, or coordinates, is well known from geography, where a location on the earth's surface is identified by latitude and longitude. It is a great circle (see Figure 2) with a fixed point F on it and its two poles, P_1 and P_2. The coordinates of a point S on the sphere's surface are now found in the following fashion. Through the poles P_1 and P_2 and the point S is drawn a great circle that intersects the basic great circle at the point S' (of the two possibilities, S' is the one for which the arc $S'S$ is less than 90°). The two coordinates (x, y) of S are then simply

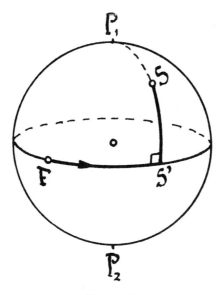

Figure 2.

$$x = FS',$$
$$y = S'S,$$

where both arcs are measured in degrees. To avoid ambiguity in the measure of the arc FS', the basic great circle is usually provided with a positive direction and FS' can then assume values up to $360°$. One may, as an alternative, choose to keep the values of FS' below $180°$ and then indicate which way the arc is measured [this happens for terrestrial longitudes given in degrees ($<180°$) east or west of Greenwich]. For the other coordinate, the arc $S'S$, already less than $90°$, one must indicate whether S lies toward P_1 or P_2 from S'. This may be done by a sign; say, positive toward P_1 and negative toward P_2, or by words such as north and south (as for terrestrial latitudes).

I shall mention here three coordinate systems on the celestial sphere. In the first the basic great circle is the ecliptic, the fixed point F is the vernal equinox (i.e., one of the points of intersection between the equator and the ecliptic), and the positive direction on the ecliptic is that of direct motion (i.e., toward the east). The coordinate measured on the ecliptic is called *celestial longitude*, or simply *longitude*, when no confusion with terrestrial longitude is likely. It is usually designated with the letter λ. The other coordinate, measured perpendicularly to the ecliptic, is called *celestial latitude* or simply *latitude*, with a similar caution as for longitude. It is positive for points north of the ecliptic, and negative for the other hemisphere. It is usually denoted by the letter β.

This ecliptic coordinate system is without question the most important in ancient astronomy; it is still used for problems involving sun, moon, and planets, for they are all on or near the ecliptic or, in other words, they have celestial latitudes that are always small ($\beta = +10°$ and $-10°$ are the extremal values for Venus, the planet showing the largest latitudes).

The longitude of the sun, whose latitude is always 0, increases by $360°$ in the course of a year. Direct motion of a planet means that its longitude increases; retrograde motion corresponds to decreasing longitude. The synodic arc of a planet—$\Delta\lambda$, as we have already called it—is the increment in the planet's longitude from one synodic phenomenon to the next of the same kind.

While we count longitudes from $0°$ to $360°$, the classical manner of giving longitudes is in terms of a zodiacal sign and degrees within that sign. The ecliptic in this system is divided into 12 parts of exactly 30° each, beginning with the vernal equinox, and they are named thus:

♈ Aries:	$0° \leq \lambda < 30°$		♎ Libra:	$180° \leq \lambda < 210°$
♉ Taurus:	$30° \leq \lambda < 60°$		♏ Scorpio:	$210° \leq \lambda < 240°$
♊ Gemini:	$60° \leq \lambda < 90°$		♐ Sagittarius:	$240° \leq \lambda < 270°$
♋ Cancer:	$90° \leq \lambda < 120°$		♑ Capricorn:	$270° \leq \lambda < 300°$
♌ Leo:	$120° \leq \lambda < 150°$		♒ Aquarius:	$300° \leq \lambda < 330°$
♍ Virgo:	$150° \leq \lambda < 180°$		♓ Pisces:	$330° \leq \lambda < 360°$

and I have affixed to them their now-standard sigilla. Thus, to give an example, Leo 27° is simply another way of denoting the longitude 147°.

The Babylonian zodiac does not agree exactly with this description; at the period, from which most of our texts come, the signs begin some 5° earlier than in the preceding list. We shall return to the reason for this later.

The longitude of the sun at vernal equinox is then Aries 0° or $\lambda = 0°$; at summer solstice Cancer 0°, or $\lambda = 90°$; at fall equinox Libra 0°, or $\lambda = 180°$; and at winter solstice Capricorn 0°, or $\lambda = 270°$.

I shall briefly mention two more coordinate systems. The first has as its basic great circle the celestial equator whose poles are the north and south poles, and the fixed point F is again the vernal equinox. The first coordinate is called *right ascension* and is measured from the vernal equinox toward the east so that the sun's right ascension always increases. Right ascension is usually denoted by the letter α. The other coordinate, measured perpendicularly to the equator, is called *declination* and is counted positive north and negative south of the equator. Declination is usually denoted by the letter δ.

The sun's right ascension at the four cardinal positions just mentioned is, respectively, 0°, 90°, 180°, and 270°; that is, its right ascension here is the same as its longitude, which, of course, is not so elsewhere. The declination of the sun at equinox is $\delta = 0°$, at summer solstice $\delta = 23\frac{1}{2}°$, and at winter solstice $\delta = -23\frac{1}{2}°$, where the value $23\frac{1}{2}°$ is the inclination of ecliptic to equator.

In the equator and the ecliptic systems, the coordinates of fixed stars will not change appreciably in the course of even several years; that they change at all is largely due to the precession of the equinoxes, which we will describe later. The longitudes, latitudes, right ascensions, and declinations of the other celestial bodies change at a moderate rate, most swiftly for the moon, whose longitude grows by about 13° per day, while solar longitude increases about 1° per day. The daily rotation has no effect in these coordinate systems at all.

That is not so in the last system I shall introduce. Here the horizon is the basic great circle whose poles are zenith and nadir, and the fixed point F is the south point. The first coordinate is called *azimuth* and is usually counted from the south point westward from 0° to 360°. It is denoted by Az. The other coordinate, measured perpendicularly to the horizon, is called the *altitude* and is denoted by h. It is counted positive for objects above the horizon, and negative for those below.

The horizon system is a local system; it belongs to the observer and his or her terrestrial surroundings. The altitude—what I earlier called elevation—of the north pole is ϕ, the terrestrial latitude of the observer, and the azimuth of an object not too close to a pole changes about 360° in one day due to the daily rotation.

Precession, Anomalistic Periods, and Various Other Refinements

In this section I deal with various refinements—long-term changes, and variation in velocities—which I have hitherto ignored for the sake of simplicity of presentation.

The first of these is the *precession of the equinoxes*, or simply the *precession*. Its effect is most easily described in the ecliptic system: The precession causes the longitude of all fixed stars to increase at a uniform, very slow rate of 1° in approximately $72\frac{2}{3}$ years, while their latitudes remain constant.

Of course, this phenomenon can be, and often is, described in a different manner. Since longitudes are measured from the vernal equinox, one can just as well say that the vernal equinox moves in the retrograde direction—it precedes—among the fixed stars along the ecliptic—a great circle that remains fixed among the fixed stars since their latitudes do not change. Thus, the equator, which serves to define the vernal equinox, moves slowly among the fixed stars but in such a fashion that its inclination to the ecliptic remains the same; the north pole travels slowly on a small circle of radius $23\frac{1}{2}°$ and the northern pole of the ecliptic as its center.

Right ascension and declination of a fixed star will therefore both change, so it is more complicated to describe the effects of the precession in the equatorial than in the ecliptic system.

One consequence of the precession of the equinoxes is that we should distinguish between two kinds of year. One is the period of the sun's return to a fixed star, and it is called the *sidereal year*; the other is the period of the sun's return in longitude (i.e., to the vernal equinox) and, since this governs the seasons, it is called the *tropical year*. The latter is slightly shorter than the former for the sun has to travel slightly farther to catch up with a fixed star than with the vernal equinox, which moves slightly to meet it during each of its revolutions. Values of the two are

1 sidereal year = 365.256 days,

1 tropical year = 365.242 days.

Similarly, we distinguish between a *sidereal* and a *tropical* month, the periods of the moon's return to a fixed star and to the same longitude, respectively. Their values are

1 sidereal month = 27.32166 days,

1 tropical month = 27.32158 days.

I must most emphatically warn that my inclusion of precession in this presentation does not imply that it plays a role in Babylonian astronomical schemes. There is no evidence whatever that the phenomenon of precession was known to Babylonian astronomers; all indications are that their longitudes are sidereal longitudes—that their zodiac was fixed among the fixed stars, that their year was the sidereal year, and that they were unaware of any difference between this and the seasonal or tropical year. It was Hipparchus (ca. 150 B.C.) who first drew the distinction between the tropical and the sidereal year or who, in other words, discovered the precession. I introduced the precession in this presentation to make clear the natural discrepancy between Babylonian and modern longitudes, a discrepancy that increases with time at the rate of 1° in 72 to 73 years.

Anomalistic Periods

In the previous discussion of the motions of sun and moon among the fixed stars, I have been content to give crude estimates of average daily progress. In a more precise account one must recognize that the velocity—which in spherical astronomy means, as it must, apparent angular velocity—of either body is not constant, but changes periodically by sensible amounts.

The sun's daily progress varies between some $1.02°$ (or $1°1'$) and $0.95°$ (or $57'$) in a regular manner; the period of this variation, that is, the time interval from high velocity to high velocity, or from low velocity to low velocity, is called one anomalistic year. Its length is

$$1 \text{ anomalistic year} = 365.2596^d$$

(where ^{d} = days) which is very close to the value of the sidereal year. Within the historically relevant period we may ignore this difference and shall, with the Babylonian astronomers, consider solar velocity as a function of Babylonian, namely sidereal, longitude.

For the moon the situation is quite different. Its daily progress varies between some $15°$ and $11\frac{1}{2}°$ (it makes no sense to present sharper limits here), and the period from high velocity to high velocity, called the *anomalistic month*, is

$$1 \text{ anomalistic month} = 27.5545^d.$$

This is substantially more than either the sidereal or the tropical month—by some 6 hours—and the anomalistic month was recognized as a separate and important parameter in Babylonian lunar theory, as we shall see. Abandoning for the moment the conventions of spherical astronomy, I may state that the moon assumes its least velocity when at the point of its orbit farthest from the earth, or at its apogee, and its greatest velocity when nearest, or at its perigee; that the anomalistic month is longer than the tropical month means, then, that the apogee and the perigee of the moon both advance steadily in the ecliptic.

Synodic Arcs of the Planets

A quantitative investigation of the synodic arc $\Delta\lambda$ of a given planet (i.e., the increment in the planet's longitude when it moves from one synodic phenomenon to the next of the same kind) will show that it is not constant but varies according to where in the ecliptic the planet happens to be. To say this in other words, $\Delta\lambda$ is a function of λ, the longitude at which the arc $\Delta\lambda$ begins. That this is correct when longitudes are counted sidereally as in Babylonian astronomy can be established on the basis of modern planetary theory, but I shall not do so here, though the necessary arguments are fairly simple.

Long-Term Changes

There are several long-term, or exceedingly slow, changes in the various parameters I have mentioned, in the lengths of the several kinds of month, in the length of the year, in the amount of the precession, in the relative positions of fixed stars, in the places where synodic arcs achieve their maxima, and so on. These minute variations, some of which are called *secular* since they are only felt after the lapse of centuries, are of prime interest to astronomers. I shall nonetheless pass them by, for in the time interval of Babylonian astronomy, and within its margin of precision, they play no significant role. The causes of some of these very slow changes are still not quite under control; indeed, some of their magnitudes are not yet unequivocally established. Ancient observations, once properly understood and properly treated, must play a central role in these matters, but our understanding of ancient theoretical astronomy clearly does not depend on such issues at all.

1
Babylonian Arithmetical Astronomy

In late May 1857 a committee, appointed by the Royal Asiatic Society of Great Britain and Ireland, met in London to compare four independent translations of an Assyrian text inscribed in cuneiform characters in duplicate on two well-preserved clay cylinders. Hormuzd Rassam had found them as foundation deposits in the ruins of ancient Ashur in 1853 when he was digging in Mesopotamia on behalf of the British Museum. W. H. Fox Talbot, the gentleman scientist, inventor of photography, and linguist, had been given a copy of the text by H. C. Rawlinson, the remarkable soldier, diplomat, and linguist, and sent his sealed translation of it to the Society with the suggestion that other scholars be invited to translate the same text so the results could be compared to test the validity of the decipherment of Assyrian for, as he writes, "Many people have hitherto refused to believe in the truth of the system by which Dr. Hincks and Sir H. Rawlinson have interpreted Assyrian writings, because it contains many things entirely contrary to their preconceived opinions."*

The Society chose the Reverend Dr E. Hincks, Ireland, Lieut.-Col. Sir Henry C. Rawlinson, London, and the Orientalist Dr. J. Oppert, who happened to be in London on leave from Paris. The three sent in their translations of the text—the annals of the Assyrian king Tiglath-Pileser I (11th century B.C.) as it turned

---

*The committee's reports and the four translations are published in the pamphlet: "Inscription of Tiglath Pileser I., King of Assyria, B.C. 1150, as translated by Sir Henry Rawlinson, Fox Talbot, Esq., Dr. Hincks, and Dr. Oppert." Published by the Royal Asiatic Society. London, 1857. And again in: Journal of the Royal Asiatic Society, Vol. XVIII, 1861, pp. 150–219.

out to be—and the appointed committee opened and compared them, including Fox Talbot's. In the words of a report dated May 29, 1857, "Having gone through this comparison, the Examiners certify that the coincidences between the translations, both as to the general sense and verbal rendering, were very remarkable."*

This dramatic event marks the end of the heroic epoch of the decipherment of cuneiform script by the great amateurs, as they had to be, and Assyriology had been established as a scholarly discipline. Much hard and patient work with the cuneiform texts lay ahead before Akkadian, as the Semitic language of the texts is called, took its place beside Hebrew, Greek, Latin, and Egyptian as one of the well-understood classical languages, but the correct path had been found.

The texts are clay tablets, almost all unbaked, inscribed with signs, each of which is a cluster of wedges impressed in the wet clay with a sharpened stylus—"cuneiform" means "wedge-shaped." The difficulties with the system of Hincks and Rawlinson, as Fox Talbot called it, were two: They had been forced to assume, first, that most signs could be read either as an ideogram—an entire word—or as a syllable and, second, that in neither case was there a one-to-one correspondence between signs and readings. I shall give but one example. The sign ⟨⟩ is usually transcribed in the astronomical texts as ge_6. As an ideogram it may have one of the obviously related meanings: to be black, black, and night. In Akkadian it would then be pronounced ṣalâmu (or declined form), ṣalmu, and mûšu, respectively. However, as a syllable it is sixth in a list of 27 signs, each of which can be pronounced gi or ge, and it may also have the phonetic values mi, me, and ṣil. When one further realizes that there are more than 500 distinct signs, one can readily perceive the formidable obstacles both to the execution of the decipherment and to its acceptance, but Hincks and Rawlinson were right.

Three groups of texts will be of particular interest in what follows. The first came to the British Museum in the early 1850s from excavations in Kuyunjik—the ancient Ninive—in two lots. The first was found by Layard in the southwest palace, and the second by Rassam in the north palace, in the room, incidentally, that was decorated with the famous reliefs of Aššurbanipal's lion hunt. Together they form the royal library—or Aššurbanipal's library, as it is often called. It was a stroke of luck that this large body of texts came to light this early, for it contains, often in standard Assyrian script, many copies of much older texts of many varieties—Aššurbanipal (668–627 B.C.) was an avid collector. The bulk of the texts centered on the series *Enūma Anu Enlil* of celestial omens comes from this library.

Second, in the last decades of the last century, texts began arriving in the British Museum by the tens of thousands. They were bought from dealers who, in turn, obtained them through unscientific excavation. Among them is a group of some 1500 texts that come from what must have been an extensive astronomical archive somewhere in the city of Babylon. Approximately 1200 of these are nonmathematical astronomical texts, classified by A. Sachs as diaries, almanacs, and goal-year texts. The rest are concerned with mathematical astronomy, and I shall refer to them as being of the "*ACT* type" after the standard abbreviation for O. Neugebauer's *Astronomical Cuneiform Texts*, where most of them were published. I should add that quite a few texts from the astronomical archive found their way to other collections in Europe and America.

Last, there is a smaller group of texts, mostly of the *ACT* type, that were found in Uruk, very likely within the domain of the Reš sanctuary. The largest single set of fragments from this site is in the Archaeological Museum in Istanbul—Iraq was part of the Turkish Empire before the first World War—but many others, among them numerous potential joins with Istanbul fragments, leaked from the excavations and ended up in many other museums.

In respect of content we have, then, three principal classes of astronomical texts. First there are those that in one way or another are concerned with astronomical omens. At the core of this class is the series of about 70 tablets now, as in antiquity, known by its *incipit Enūma Anu Enlil* ["when (the gods) Anu and Enlil"]. The contents of this collection of omens are presumably very old, but only a few Old-Babylonian (i.e., from the early second millennium B.C.) fragments of some of the tablets are known. The two-tablet series Mul Apin is more astronomical in character and very likely younger than *Enūma Anu Enlil*. Finally, there are the royal reports written by specialists in astronomical omens from various cities of the kingdom to the Assyrian king. Such a report may contain an observation and an interpretation of its significance according to the canonical texts, mostly *Enūma Anu Enlil*. These reports date from about 700 B.C.

The second class, that of the astronomical diaries and related texts, contains or is based on a high proportion of observations. They come from the astronomical archive in Babylon and span in time the interval from about 750 B.C. to A.D. 75—the text from this last year happens to be the latest datable text written in cuneiform.

Third, the mathematical astronomical texts, those of the *ACT* type, come from astronomical archives in Babylon and from Uruk. They are, in respect of date, mainly, from the last four or five centuries B.C.—the Uruk texts stop already around 150 B.C. —so they represent one of the last, as well as one of the finest, contributions of Mesopotamian culture.

In the following sketch of Babylonian astronomy, I shall concentrate on the third kind of texts, but the two groups of non-mathematical texts cannot well be completely excluded from a discussion of Babylonian astronomy, for they are connected both to each other and to the *ACT* texts. Indeed, all three categories of texts were kept in the archive in Babylon; the scribes who wrote even the elaborate theoretically computed ephemerides of *ACT* called themselves by the title *ṭupšar Enūma Anu Enlil* ["scribe of (the series) *Enūma Anu Enlil*"]; the diaries can be viewed as collections of raw material for omens, and they provided in the process the observational basis for constructing the theories behind the *ACT* texts; and, finally, the *ACT* texts predict precisely the core of the celestial phenomena recorded in the diaries.

Last, a few remarks about the astronomical archive in Babylon. Colophons, names of scribes, scribal families, runs of the British Museum's accession numbers, and many other features ensure its existence, but we do not know its precise location within the vast ruins of Babylon, for all the texts from it were excavated without records. The earliest diary is dated by A. Sachs to −651, and he published six texts from A.D. 31 to A.D. 75. These late texts throw a particularly interesting light on the activities around the archive. Already in 275 B.C. the government had moved from Babylon to Seleucia and Antiochus ended Babylon's civil existence. The once great and glamorous metropolis fell rapidly into decay, and some three centuries later Strabo, who died circa A.D. 20, writes that the greater part of Babylon is deserted and quotes the comic poet who said, "The Great City is a great desert." Yet these six texts show that even after Strabo's time there were still people living in the ruins of

Babylon who not only knew, and taught others, the difficult art of reading and writing technical Akkadian in cuneiform, but who also had access to the astronomical archive and the desire and competence to use and increase it. Here one may well recall that Pliny (d. A.D. 79) in *Nat. Hist. VI*, xxx, 121–122, remarks about Babylon, "The temple of Jupiter Belus still remains— it was here the creator of the science of astronomy was—the rest has reverted to desert." Indeed, many of the astronomical texts from Babylon carry the invocation *ina amat Bel u Beltia lišlim* ["at the command of (the deities) Bel and Belti, may it go well"].

The study of astronomical cuneiform texts began already in the 1870s. The Orientalist P. Johann N. Strassmaier, S. J., was then engaged in classifying and copying tablets in the British Museum's collection, which was growing enormously in precisely those years. In the matter of texts that he deemed astronomical he sought the advice of his fellow-Jesuit, the mathematician and astronomer P. Joseph Epping.

Epping brilliantly and swiftly came to understand much of the basic structure and terminology of Babylonian astronomy, and he succeeded in correlating precisely Seleucid and Western chronology. He summed up these early researches in a charming little book, *Astronomisches aus Babylon* (1890), which still conveys a sense of the excitement of his pioneering efforts.

After Epping's death in 1894, his work was continued by P. Franz Xavier Kugler, S. J., at first also in collaboration with Strassmaier. Kugler addressed his task with rare vigor and imagination, and he published his many results in volumes entitled *Die babylonische Mondrechnung* (1900) and *Sternkunde und Sterndienst in Babel* I (1907) and II (1909–1924) with three supplements, the last of which Fr. Johann Schaumberger wrote in 1935.

In his editions Kugler included astronomical texts of all kinds. In the mid-1930s Otto Neugebauer selected those dealing with mathematical astronomy to subject them to a systematic and uniform treatment. Later he was given a copy of Strassmaier's notebooks, which led to the identification of a large number of new texts in the British Museum, and many texts from other collections—those in Istanbul, Chicago, the Louvre, Yale, and Berlin among them—came to his notice, so it was some 20 years before he published his *Astronomical Cuneiform Texts (ACT)*

(1955). It contains editions of all the texts of his chosen kind known at that time, about 300 in number. Neugebauer included for the sake of completeness some 50 previously published texts, many of them augmented by joins.

Shortly before the appearance of *ACT*, A. Sachs got the opportunity to survey part of the collection of tablets in the British Museum in order to identify astronomical texts. The result was overwhelming. Before his search Sachs had been aware of 12 nonmathematical astronomical texts—astronomical diaries and related texts—but now he had 1200, and they are still in the process of publication. Further, Sachs was given a large number of beautiful hand copies executed by the remarkable T. G. Pinches. They had lain unnoticed in the British Museum for half a century until Sachs published them with his comments and datings as *Late-Babylonian Astronomical and Related Texts Copied by T.G. Pinches and J.N. Strassmaier* (*LBAT*). Finally, during his survey Sachs discovered some 200 texts concerning mathematical astronomy. Many of these have been analyzed and published in various journals, mostly by Sachs, Neugebauer, and myself. Among the contributors to our further and deeper understanding of already published texts, I must particularly mention B. L. van der Waerden, A. Pannekoek, P. Huber, J. Britton, and N. T. Hamilton.

General Background

Number System

Babylonian mathematics appears fullblown in Old-Babylonian texts (from the early second millennium B.C.) and changes little, if any, afterward. It deals principally with problems that we today would classify as belonging to arithmetic, number theory, and elementary algebra, although it also deals with certain geometrical problems, some of them involving the use of the Pythagorean theorem. It is based on a most efficient means of writing numbers, a place-value system of base 60—thus we call it *sexagesimal*—with many similarities to our decimal system.

The single digits are written as combinations of two signs: a vertical wedge meaning 1, and a corner wedge meaning 10. The numbers from 1 to 59 are written thus:

𒐕 𒐖 𒐗 𒐘 . 𒐚 . . 𒐛 𒌋 𒌋𒐕 . . 𒎙 . 𒌍 . . 𒐏 . . 𒐐 . . 𒐎

1 2 3 4 .7 . .9 10 11 . . 20 . .30 . .40 . .50 . . 59

(there is a late cursive form for 9 consisting of three diagonal wedges).

The sexagesimal system uses these digits to express all whole numbers, as well as certain fractions, by assigning importance to a digit's place within a sequence of digits: Moving a digit one place to the left increases its value 60 times, and moving it one place to the right means to divide by 60, even beyond the units' place. When we transcribe Babylonian numbers, we separate the digits by commas so 1,25,30 can mean

$$1 \cdot 60^2 + 25 \cdot 60 + 30 = 5130.$$

However, the ancients did not indicate the units' place, nor did they employ terminal zeros, so this sequence of digits could, in principle, stand for 5130 multiplied by any power of 60, positive or negative. In practice, there is rarely any doubt about the absolute size of such a number, and when from context we have determined the units' place, we separate the whole from the fractional part with a semicolon (analogous to our decimal point) so that

$$1,25;30 = 1 \cdot 60 + 25 + \frac{30}{60} = 85\frac{1}{2}.$$

I should repeat for emphasis that the semicolon has no equivalent in the texts but is solely introduced for the sake of convenience in our translations.

In the late astronomical texts the separation sign 𒀹 (which we transcribe as ".") is used to indicate an empty sexagesimal place, much like our zero, and also to help the reader distinguish, for example, between 10,3 (written 10,.3) and 13.

Calendar

The Babylonian calendar is strictly lunar: A new month begins on the evening when the crescent of the new moon becomes visible for the first time. A month contains either 29 days (hollow) or 30 days (full); on the average its length is the mean synodic month, which is slightly more than $29\frac{1}{2}$ days. In the long run there are then slightly more full than hollow months. I should emphasize that the sequence of full and hollow months shows no simple pattern. The *ACT* material contains several instances of four consecutive full months; in one of the years in question we also find three consecutive hollow months (*ACT*, p. 94). Indeed, it is a very complicated matter to predict the visibility of the new moon, and it is one of the great triumphs of Babylonian lunar theory that it succeeded in doing just that so well.

A normal year consists of 12 lunar months, but since they amount to only some 354 days, an extra month was occasionally introduced in order to keep the year in step with the seasons. From about the mid-fifth century B.C., these intercalations followed a rigid scheme based on the so-called Metonic cycle, which equates 19 years with 235 months. Every group of 19 years contained seven years with intercalary months, six of them inserted as a second month XII (we denote it XII_2) and one as a second month VI (VI_2), and the pattern is fixed modulo 19 years. In our transcriptions we usually mark a 13-month year containing an XII_2 with a single asterisk, and with a double asterisk a year with a VI_2. The Babylonian year began near, and mostly after, vernal equinox.

The Akkadian names of the months, and the late texts' ideograms for them, are as follows:

I	Nisannu	bar	VII	Tašrītu	du_6
II	Aiaru	gu_4	VIII	Araḫsamna	apin
III	Simānu	sig	IX	Kislīmu	gan
IV	Dūzu	šu	X	Ṭebētu	ab
V	Ābu	izi	XI	Šabaṭu	zíz
VI	Ulūlu	kin	XII	Adaru	še
VI_2	kin-2-kám,	kin a	XII_2	dir-še,	dir, *a*

In earlier times a date is given by regnal year of the reigning king, month, and day, but in 311 B.C., with the official reign of Seleucus, once Alexander's general, began what became a continuing count of years. The Seleucid Era (S.E.), as it is called, remained in use throughout the *ACT* material.

The continuing year-count of the Seleucid Era and the fixed intercalation scheme are, of course, very useful when one wishes to make astronomical predictions. However, I cannot tell how one in pre-Seleucid times would indicate a date far in the future for I do not know of a single text that gives a regnal year number in excess of the natural reign of the king. Texts that date correctly in the reigns of several kings can obviously not have been written in advance of the events they describe, at least not in their entirety.

Units

In a certain sense the basic unit for measuring time in Babylonian astronomy is the synodic month, and this is intimately connected with the calendar. For shorter time intervals the day is employed, divided into 360 time-degrees (uš).

In the planetary theory we find employed an artificial "day", which is one-thirtieth of a synodic month; we call it lunar day or *tithi* (τ), for in modern times it was first encountered under that name in Indian astronomy. The terminology of the available texts draws no distinction between "day" and "*tithi*"; both are indicated by u_4 or me.

Babylonian astronomy is principally concerned with the moon, the sun, and the planets, all on or close to the ecliptic, so it is not surprising that the most important celestial coordinate is (celestial) longitude, λ. In the late texts we find the ecliptic divided into 12 zodiacal signs, each of length precisely 30 degrees (uš)—our modern degree is of Babylonian origin. The standard sigilla and Latin names of the signs, and the corresponding ideograms used in the texts are, in order,

♈	Aries	ḫun, lu	♎	Libra	rín
♉	Taurus	múl	♏	Scorpio	gír-tab, gír
♊	Gemini	maš-maš, maš	♐	Sagittarius	pa
♋	Cancer	alla$_x$ (KUŠÚ)	♑	Capricorn	máš
♌	Leo	a	♒	Aquarius	gu
♍	Virgo	absin, absin$_o$ (KI)	♓	Pisces	zib-me, zib

The zodiacal signs are, of course, derived from constellations near the ecliptic. We are not sure when the important step of replacing constellations by 30° segments was taken—it may have been in the fifth century B.C. The normalized Babylonian zodiac is sidereally fixed; thus, the systematic difference between positions computed according to Babylonian and modern theories changes with time at the rate of precession.

The degree is the standard Babylonian unit for expressing

Celestial Omen Texts

An omen consists of two parts: a *protasis* and an *apodosis*. Thus, the first of in all 59 omens in the Venus Tablet of Ammiṣaduqa (Tablet 63 of the series *Enūma Anu Enlil*) reads

In month XI, the 15th day, Venus disappeared in the west, it stayed away 3 days, and in month XI, the 18th day, Venus became visible in the east: springs will open, Adad will bring his rain, Ea his floods; king will send messages of reconciliation to king.

Here the protasis is written in a past tense, the apodosis in a future tense. It is reasonable to assume that the entire omen would become a historical statement if the apodosis were changed to the past tense. This assumption underlies the role the Venus text has played in Old-Babylonian chronology. Kugler realized that the tenth omen is incomplete and that instead of an apodosis it contains what he brilliantly read as the name of the eighth year of Ammiṣaduqa's reign, *Year of the Golden Throne*. The omens in Sections I and III of Tablet 63 were consequently taken to contain an observational record for the 21 years of Ammiṣaduqa's reign and served to limit, on astronomical grounds, the absolute date of the beginning of Hammurapi's dynasty to a few possibilities in the early second millennium B.C. The sequence of full and hollow months is highly irregular, and so contains much information. By ingenious and sophisticated use of this information, in addition to the Venus data, P. Huber* has convincingly shown that only the "Long Chronology" with Ammiṣaduqa $1 = -1701 = 1702$ B.C. makes sense, so Hammurapi began his reign in 1848 B.C., unless the Venus Tablet is altogether rejected as chronological evidence.

Section II is a highly schematic interlude based on a crude subdivision of Venus's mean synodic period. The date of this scheme is moot. Section IV consists of a rearrangement of the omens in Sections I and III in order of months.

---

*Peter J. Huber, *Astronomical Dating of Babylon I and Ur III*. Malibu, 1982

The Venus Tablet ends with a catch line—the first line of Tablet 64—saying, "If Jupiter remains (in the sky) in the morning, enemy kings will become reconciled."

The catch line is, alas, much more typical of *Enūma Anu Enlil* than Sections I and III, for the protases of the many thousands of omens in the series are in general qualitative and contain little significant or useful astronomical information.

According to Weidner, the 70 tablets of *Enūma Anu Enlil* can be classified as follows: The first 23 concern the moon; the next 20 the sun; then follow a few tablets of meteorological omens; and the last 20 deal with planets and fixed stars.

In the lunar protases, to take but one example, there is much concern with the general appearance of the moon, for example, on the important evening of its first visibility, whether it is light or dark, which way its horns point, or whether it is surrounded by a halo. Tablets 15 to 23 deal with similar aspects of lunar eclipses. The apodoses here, as throughout the series, predict events and conditions of concern to king or country: war and peace, quality of the harvest, and bad weather, to name but a few. This sort of astrology is called *judicial* in contrast to *personal* astrology with its horoscopes for individuals. So far we have only encountered cuneiform evidence for personal astrology in texts written well after the Assyrian period.

The imperfectly understood 70 tablets of *Enūma Anu Enlil* are at the center of what at present seems a morass of related texts: extracts, commentaries, and reports. Once all of this material is brought under control, we shall very likely have a firm grasp on the state of astrology in Mesopotamia, and the role of "diviners," near the end of the Assyrian Empire, and we may also be afforded some notion of the tradition of some of the texts.

Mul Apin (the Plow Star), the series of two, or maybe three, tablets contains some omens, but it is principally a compendium of astronomical knowledge. The treatment, theoretical if you will, of various astronomical phenomena is always schematic in the extreme. Thus, there is a section relating the first appearances of certain fixed stars and constellations to an idealized calendar in which a year consists of 12 months of 30 days each. Likewise there is a part giving the length of daylight and night for the 15th day (full moon) of each month in the same calendar. This daylight scheme is based on a ratio of 2:1 between longest and shortest daylight, a ratio also found in Tablet XIV of *Enūma Anu Enlil* and in a recently published text from the mid-seventh century, though the latter text elsewhere gives the ratio as 3:2.

In Hellenistic Greek astronomy the ratio of longest-to-shortest daylight is a common way of denoting the terrestrial latitude of a place. Here, following the later Babylonian texts, the canonical ratio for Babylon (or rather for the entire Second Climate) is 3:2, while 2:1 corresponds to a parallel far north of Babylon's. I must emphasize that the cuneiform texts give no evidence whatever of an awareness that this ratio changes when you travel north or south.

In this section of *Mul Apin* the length of daylight changes from month to month by a constant amount from maximum to minimum and back again. We see here an embryonic form of the linear zigzag function that was to become one of the two major devices of expressing periodic phenomena in the later texts of the *ACT* type.

We cannot at present determine the dates of the various parts of the series *Mul Apin* nor the tradition of the texts.

The Astronomical Diaries and Related Texts

To introduce the astronomical diaries I present part of a composite diary translation from the edition by A. Sachs and H. Hunger of all extant astronomical diaries. The tablet consists of four rejoined fragments in the British Museum [Rm. 718 + Rm. 723 + BM 32840 (= 77 − 2 − 22,2) + BM 34130 (= Sp. 232)]; when unbroken it gave reports for the last six months of year 60 of the Seleucid Era (Oct. 252–Mar. 251 B.C.), but only those for months VII, VIII, and XII are preserved. My excerpt covers month VII.

Time intervals are given in time degrees ($1° = 4$ min.), and the frequently used cubit (kùš) is here $2°$, or maybe $2\frac{1}{2}°$, of arc.

No. −251

Obv.'

1 [Year 60,] king [Antioch]us. Month VII, the 1st (of which followed the 30th of the preceding month), sunset to moonset: $13°$, it was low. Night of the 2nd, the moon stood $2\frac{1}{2}$ cubits in front of Venus to the west. Night of the 3rd, the moon was $1\frac{1}{2}$ cubits in front of ϑ Ophiuchi, 2 cubits below Saturn.

2 [.... Night of the 4th (or: 5th), first part of the night, Venus was] 2 cubits [above] α Scorpii; last part of the night, Jupiter was 6 fingers above ρ Leonis. Night of the 6th, beginning of the night, the moon was [nn] cubits in front of β Capricorni. Night of the 8th, beginning of the night, the moon was

3   [(nn cubits) nn] fingers [.... δ Capricorni.] The 11th, Mercury's first appearance in the west, omitted. Night of the 12th, beginning of the night, the moon was 1 cubit in front of [η] Piscium. Night of the 13th, moonrise to sunset: 4° 10'. The 13th, moonset to sunrise: 3° 50', measured?.

4   [Night of the 14th, sunset to moonrise: 2° 20'; last part of the night, the moon was] 1 cubit behind? η Tauri. The 14th, sunrise to moonset: 10°, measured. Night of the 15th, first part of the night, Venus was 1 cubit above ϑ Ophiuchi; last part of the night, the moon was $2\frac{1}{2}$? cubits behind α Tauri. Night of the 16th, last part of the night, the moon was

5   [nn cubits behind ζ Tauri. Night] of the 17th, last part of the night, the moon was 2 cubits above γ Geminorum. Night of the 18th, last part of the night, the moon was $1?\frac{1}{2}$ cubits in front of α/β Geminorum. Night of the 20th, last part of the night, the moon was .... in front of ε Leonis. Night of the 21st, last part of the night, the moon was

6   [nn cubits .... α Leonis. Night of the 22nd, last part of the night,] the moon was $1\frac{1}{2}$ cubits in front of ϑ Leonis. Night of the 23rd, last part of the night, the moon was 1 cubit in front of β Virginis. Night of the 25th, last part of the night, the moon was 1? cubit in front of α Virginis. Around the 26th, Mercury's last appearance in the west, omitted.

7   [The 27th?, Mars'? first appearance? in Libra?; moonrise to sunri]se: [20°], measured. That month, the equivalent was: barley, 1 kur 1 pān 4 sūt, at the end of the month, 1 kur [x pān] 3 sūt 3 qa; new dates, 3 pān; mustard, 1 kur 2 pān 3 sūt; cress, 2 pān 3 sūt; sesame, 1 pān 5 sūt;

8   [wool, nn minas] for 1 shekel of wrought silver. At that time, Jupiter was in Leo; Venus was in Scorpius, at the end of the month in Sagittarius; the 11th, Mercury's first appearance in the west, omitted; the 26th, Mercury's last appearance in the west, omitted;

9   [Saturn was in Scorpius, at the end of the month in Sagittarius; the 2]7th?, Mars' first appearance in Libra. That month, the river level rose 4 fingers, total: 17 was the *na* (gauge).

A standard diary is built up of six or seven units like the preceding excerpt, each of which spans one lunar month. The principal concern is with the behavior of the moon. Six phenomena and time intervals are recorded every month. First we find a statement about the first visibility of the new crescent, for this phenomenon marks the beginning of a new month. The diary tells us whether the previous month was full or hollow and gives the time interval from sunset to moonset (na). Near the middle of the month, when the moon is close to full, four time intervals are recorded:

šú: time from moonset to sunrise
na: time from sunrise to moonset
me: time from moonrise to sunset
ge$_6$: time from sunset to moonrise

Near month's end the last visibility of the moon is noted as well as the time from moonrise to sunrise (kur). Otherwise the progress of the moon during the month is recorded, particularly its passage above or below one of the 31 standard reference stars given in Table 1. These "Normal Stars," as they are also called, all lie less than 10° from the ecliptic. The curious gap of more than 60° in the signs Aquarius and Pisces still wants explanation, particularly since there are several fixed stars in this region brighter than some of the normal stars included in the list.

The diaries include two kinds of information about the five planets. First, it is noted when a planet passes above or below a Normal Star, and by what amount; "above" and "below" probably refer to ecliptic coordinates. Second, the date of each planetary phase is recorded together with the zodiacal sign in which the planet is at that time or, for stations, with distance to a Normal Star. The phases, or synodic phenomena, of an outer planet are

Γ: first appearance
Φ: first stationary point
Θ: "opposition"
Ψ: second stationary point
Ω: disappearance

and for an inner planet

Γ: first visibility ⎫
Φ: stationary point ⎬ in the east (morning star)
Σ: disappearance ⎭
Ξ: first visibility ⎫
Ψ: stationary point ⎬ in the west (evening star)
Ω: disappearance ⎭

where the planet is retrograde from Ψ to Φ.

The study of planetary phases is fundamental to Babylonian planetary theory, and it should be noted already here that the

TABLE 1. The standard Babylonian reference stars as they appear in texts after approximately −300

Babylonian name	Star	−600 λ	−600 β	−300 λ	−300 β	0 λ	0 β
múl kur šá dur *nu-nu*	η Pisc	350.7°	+5.2°	354.9°	+5.2°	359.0°	+5.3°
múl igi šá sag ḫun	β Arie	357.9	+8.4	2.0	+8.4	6.2	+8.4
múl ár šá sag ḫun	α Arie	1.5	+9.9	5.7	+9.9	9.8	+9.9
múl-múl	η Taur	23.9	+3.8	28.0	+3.8	32.2	+3.8
is da	α Taur	33.7	−5.7	37.8	−5.6	42.0	−5.6
ŠUR gigir šá si	β Taur	46.5	+5.2	50.6	+5.2	54.8	+5.2
ŠUR gigir šá u$_x$	ζ Taur	48.7	−2.5	52.8	−2.5	57.0	−2.5
múl igi šá *še-pít* maš-maš	η Gemi	57.4	−1.2	61.5	−1.2	65.7	−1.1
múl ár šá *še-pít* maš-maš	μ Gemi	59.2	−1.1	63.3	−1.1	67.5	−1.0
maš-maš šá sipa	γ Gemi	63.0	−7.1	67.1	−7.0	71.3	−7.0
maš-maš igi	α Gemi	74.2	+9.9	78.4	+9.9	82.5	+9.9
maš-maš ár	β Gemi	77.5	+6.5	81.6	+6.5	85.7	+6.5
múl igi šá alla$_x$ šá u$_x$	θ Canc	89.7	−1.0	93.8	−1.0	98.0	−0.9
múl igi šá alla$_x$ šá si	γ Canc	91.5	+3.0	95.6	+3.0	99.8	+3.0
múl ár šá alla$_x$ šá u$_x$	δ Canc	92.6	0.0	96.7	0.0	100.9	0.0
sag A	ε Leon	104.6	+9.5	108.7	+9.5	112.9	+9.6
lugal	α Leon	113.9	+0.4	118.0	+0.4	122.2	+0.4
múl tur šá 4 kùš ár lugal	ρ Leon	120.3	0.0	124.4	0.0	128.6	0.1
GIŠ-KUN A	θ Leon	127.3	+9.7	131.4	+9.7	135.6	+9.7
gìr ár šá A	β Virg	140.5	+0.6	144.7	+0.7	148.9	+0.7
dele šá igi absin	γ Virg	154.4	+3.0	158.5	+3.0	162.6	+3.0
sa$_4$ šá absin	α Virg	167.8	−1.9	171.9	−1.9	176.1	−1.9
rín šá u$_x$	α Libr	189.0	+0.7	193.2	+0.6	197.3	+0.6
rín šá si	β Libr	193.3	+8.8	197.4	+8.8	201.6	+8.7
múl múrub šá sag gír-tab	δ Scor	206.5	−1.7	210.6	−1.7	214.8	−1.7
múl e šá sag gír-tab	β Scor	207.1	+1.3	211.2	+1.3	215.4	+1.3
si$_4$	α Scor	213.7	−4.2	217.8	−4.3	222.0	−4.3
múl kur šá kir$_4$ šil pa	θ Ophi	225.3	−1.5	229.4	−1.5	233.6	−1.6
si máš	β Capr	267.9	+4.9	272.1	+4.9	276.2	+4.8
múl igi šá suḫur-máš	γ Capr	285.6	−2.3	289.7	−2.3	293.9	−2.4
múl ár šá suḫur-máš	δ Capr	287.3	−2.1	291.5	−2.2	295.6	−2.2

computed ephemerides predict with considerable accuracy not only dates but also planetary longitudes of the phases. Yet we saw that the diaries mention nothing but the zodiacal sign of a planet at phase, that is, its position is given only within an interval of length 30°. It has been shown how theoretical schemes of such excellence may be derived from a set of such crude observations, for I do not doubt that the diaries provided the observational basis for Babylonian mathematical astronomy. The argument is, in principle, that the observations make up in quantity for what they want in quality (Aaboe [1980]).

The diaries mention whenever an eclipse is seen at Babylon. The text BM 45745 offers a spectacular example in its report on the solar eclipse of April 15, 136 B.C., total at Babylon, very likely the best account of a solar eclipse from antiquity:

[Year 175 (of the Seleucid Era), month XII_2...] Daytime of the 29th, 24° after sunrise, a solar eclipse beginning on the south west side [...] Venus, Mercury and Normal Stars were visible; Jupiter and Mars, which were in their period of invisibility, were visible in that eclipse [...] (the shadow) moved from south west to north east. (Time interval of) 35° from obscuration to clearing up (of the eclipse). In that eclipse the north wind which [...].

A group of texts, falling into several classes, derive from the diaries. Thus, we find extracts of information, either about a particular planet or, for the moon, about eclipses, usually for an astronomically significant period. The "Goal-Year Texts," in Sachs's terminology, offer predictions for a certain year— the goal year—about the moon and planets gleaned from older diaries that antedate the goal year by appropriate periods. In the standard order of the texts, these periods are for

Jupiter: 71 years and 83 years   Saturn: 59 years
Venus: 8 years   Mars: 79 years and 47 years
Mercury: 46 years   Moon: 18 years

Of the two periods for Mars and for Jupiter, the first is used for prediction of synodic phenomena and the other for the planet's return to Normal Stars.

This method of predicting astronomical events is quite efficient, which in part may explain that it survived side by side with the more sophisticated theoretical schemes. As we shall see, the recognition of the periodic character of many astronomical phenomena and good values of the relevant periods lie at the very base of both lunar and planetary theories of the *ACT* texts.

The rest of the diaries' information, meteorological, economical, and historical, is not of direct concern to us here. I should point out, however, that the diaries occupy a unique position among documents of relevance to the study of ancient history. The everpresence of the swiftly moving moon enables us to date the texts, if we can date them at all, to the very day, and in sheer bulk, continuity, detail, and kind of information they are unmatched.

We now recognize that several passages in Ptolemy's *Almagest* refer to this group of texts. Two examples will suffice. First, Ptolemy chooses the first year (747 B.C.) of the reign of Nabonassar, king of Babylon, as epoch for his tables because "from that time and on the old observations have been preserved, in the large, until the present day" (*Alm.* III, 7). In the second passage he deplores that the older sustained sequences of planetary observations are mostly concerned with stationary points and heliacal risings and settings (*Alm.* IX, 2). We cannot tell in what form Babylonian observations reached Ptolemy and Hipparchus before him.

Babylonian Arithmetical Astronomy

When Neugebauer's *Astronomical Cuneiform Texts* (*ACT*) appeared in 1955, it presented in transcription and translation, and with penetrating commentary, all the then-known texts concerned with mathematical astronomy, about 300 in number. All derive from Uruk or Babylon and were likely written during the last three or four centuries B.C. Roughly half of them are concerned with the moon, and half with the five naked-eye planets.

We divide these texts according to content into ephemerides and procedure texts. An ephemeris gives, among other information, computed dates and longitudes for a sequence of interesting phenomena of a planet or the moon, not necessarily—in fact, rarely—day by day, as the name might suggest. The procedure texts present rules for calculating ephemerides.

Ideally, a complete set of procedure texts and a set of initial values should enable us to compute an ephemeris, but the road to our present understanding of Babylonian theoretical astronomy has meandered from ephemerides to procedure texts and back again. This is not only because the preserved procedure texts are fragmentary, but also because they are written in technically specific and very abbreviated terminology that in many cases could be decoded only when the relevant rules had been recovered from an analysis of ephemerides.

The texts are classified as belonging to System A or System B according to the mathematical techniques they employ. Lunar texts of System A come primarily, but not exclusively, from Babylon, and those of System B mostly from Uruk. I begin the following discussion by giving precise meaning to the distinction between the two systems.

Babylonian mathematical astronomy has two features that seem strange to modern eyes, and it may thus be in order to mention them here. First, it is entirely arithmetical in character or, in negative terms, there is no trace of geometrical models like the ones we have been accustomed to since the time of Eudoxos. Second, the cuneiform literature nowhere attempts to justify the precepts of the procedure texts; thus it has rested with modern scholars to uncover the underlying theoretical structures.

The space allotted here does not permit me to present all of Babylonian mathematical astronomy in the detail it deserves, and I have chosen to emphasize the planetary theories at the expense of lunar theory, contrary to their relative merit and interest. Indeed, I prefer treating one part thoroughly to treating the whole lightly, and planetary theory is by far the simpler part. For lunar theory I merely aim at conveying a sense of its complexity and sophistication, but not of its beauty, for it rests in the subtle and elegant interplay of its technical details. Even so, the planetary schemes serve to illustrate most of the characteristic approaches and techniques of Babylonian mathematical astronomy.

Planetary Theory

Babylonian and modern planetary theories differ in the very formulation of their central problems. Since the rise of Hellenistic mathematical astronomy, we have wanted our theories to enable us to answer the question: Given the time, where is the planet? Thus, we consider time the independent variable and seek means of deriving all other information from it.

The Babylonian approach is entirely different. First, all interest, at least primarily, is focused upon the planet only when it is at one of its characteristic synodic situations or phases: For an outer planet they are first appearance (Γ), first stationary point (Φ), opposition (Θ), second stationary point (Ψ), disappearance (Ω) (see Figure 1, which represents a run of Saturn with latitude exaggerated four times).

The next bold simplification is that we disregard all but one of

42   1. Babylonian Arithmetical Astronomy

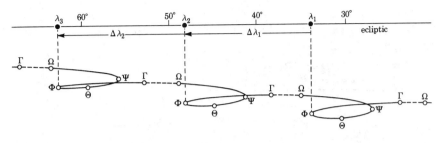

FIGURE 1.

these, say the first stationary point Φ, and we now ask the question: If we are given the longitude and the time at which a certain planet happens to be at a first stationary point, where and when will it next be at a first stationary point? What we, in the Babylonian mode, consider and wish to reproduce is then, a sequence of discrete points, in time and longitude.

The problem of finding a planet's longitude for a given date is solved—if it is addressed at all—by interpolation between the two nearest phases, as we shall see.

Planetary Theory: System A

In Table 2, columns I–IV, I have excerpted the first 25 of in all 56 lines of the text *ACT* no. 600, an ephemeris for Jupiter at first stationary point, Φ, for the years S.E. 113 to 173 (−198 to −138). According to its colophon, the text was written in Uruk S.E. 118, VII 12 (−193, Oct. 5), thus at the beginning of the time interval of its contents, so it is mostly a forecast.

In columns I and III we find year, month, and date. The year number (S.E.) is written [e.g., 1-me 13 (1 hundred 13)], but all other numbers are sexagesimals. The dates are in *tithis* and fractions thereof; the fractions are solely of computational interest. The use of the *tithi* (one-thirtieth of a synodic month) frees us from concern about which months in the future will be full or hollow; the date in *tithis* is, of course, always close to the date in days. The years with a single asterisk (the text has "a") contain a month XII_2, those with a double asterisk ("kin a") a month VI_2. The fixed 19-year pattern of intercalations of the late period can readily be established from this short excerpt.

Column II contains the difference in *tithis* ($\Delta\tau$) of the dates minus 12 months. The convention of the text is that the differ-

TABLE 2

l.	I y. SE	II Δτ	III mo. & date	IV λ	V Δλ	VI Jul. date	VII λ	VIII date
1	113**	48; 5,10	I 28;41,40 ♃	♑ 8;6 Φ		−198 Apr 22	♑ 3	II 4
	114	48; 5,10	II 16;46,50 ♃	♒14;6 Φ	36	−197 May29	♒ 8	II 21
	115*	48; 5,10	IV 4;52	♓20;6 Φ	36	−196 Jul 3	♓14	IV 12
	116	48; 5,10	IV 22;57,10	♈26;6	36	−195 Aug 9	♈21	IV 29
5	117	48; 5,10	VI 11; 2,20	♊ 2;6	36	−194 Sep 15	♉26	VI 15
	118*	45;54,10	VII 26;56,30	♋ 5;55	33;49	−193 Oct 19	♊29	VII 28
	119	42; 5,10	VIII 9; 1,40	♌ 5;55	30	−192 Nov 18	♌ 0	VIII 12
	120	42; 5,10	IX 21; 6,50	♍ 5;55	30	−191 Dec 20	♍ 1	IX 24
	121*	42; 5,10	XI 3;12	♎ 5;55	30	−189 Jan 18	♎ 1	XI 7
10	122	42; 5,10	XI 15;17,10	♏ 5;55	30	−188 Feb 18	♏ 2	XI 18
	123*	43;16,10	XII 28;33,20	♐ 7;6	31;11	−187 Mar 21	♐ 4	XII₂ 4
	125	48; 5,10	I 16;38,30	♑13;6	36	−186 Apr 28	♑ 8	I 22
	126*	48; 5,10	III 4;43,40	♒19;6	36	−185 Jun 3	♒13	III 10
	127	48; 5,10	III 22;48,50	♓25;6	36	−184 Jul 9	♓19	III 30
15	128	48; 5,10	V 10;54	♉ 1;6	36	−183 Aug 15	♈26	V 17
	129*	48; 5,10	VI 28;59,10	♊ 7;6	36	−182 Sep 21	♊ 0	VII 3
	130	45; 4,10	VII 14; 3,20	♋10;5	32;59	−181 Oct 24	♋ 3	VII 15
	131	42; 5,10	VIII 26; 8,30	♌10;5	30	−180 Nov 23	♌ 5	VIII 28
	132**	42; 5,10	IX 8;13,40	♍10;5	30	−179 Dec 23	♍ 5	IX 11
20	133	42; 5,10	X 20;18,50	♎10;5	30	−177 Jan 23	♎ 5	X 22
	134*	42; 5,10	XII 2;24	♏10;5	30	−176 Feb 22	♏ 6	XII 6
	135	44; 6,10	XII 16;30,10	♐12;6	32;1	−175 Mar 26	♐ 9	XII 21
	137*	48; 5,10	II 4;35,20	♑18;6	36	−174 May 3	♑13	II 10
	138	48; 5,10	II 22;40,30	♒24;6	36	−173 Jun 8	♒18	II 28
25	139	48; 5,10	IV 10;45,40	♓30;6	36	−172 Jul 14	♓25	IV 10

ence between the date in line n and that in line $n-1$ is 12 months plus the $\Delta\tau$ listed in line n.

Column IV gives the longitude, λ, of Jupiter at first stationary point in terms of degrees of a zodiacal sign. In column V, which is not in the text, I have presented the differences, $\Delta\lambda$, of these longitudes. We note that columns II and V run parallel, so that in each line

$$\Delta\tau - \Delta\lambda = 12;5,10.$$

Thus, we need bring but one of these columns under control to uncover the structure of the text; as we shall see, there are good reasons to begin with $\Delta\lambda$.

Column V, the total progress of Jupiter from one first stationary point to the next, falls conspicuously into constant stretches of either 36° or 30° separated by intermediate values. The key to the structure of this column is the realization that the scheme is tied to the ecliptic. As a procedure text would have it (e.g., No. 821):

From Gemini 25° to Scorpio 30° add 30°. Whatever exceeds Scorpio 30°, multiply it by 1;12 and add it to Scorpio 30°.
From Scorpio 30° to Gemini 25° add 36°. Whatever exceeds Gemini 25°, multiply it by 0;50 and add it to Gemini 25°.

Thus, the ecliptic is divided into two parts—the fast and the slow arcs—inside which the phenomenon progresses in steps of 36° and 30°, respectively. If such a step crosses a boundary of the arcs, the amount that reaches into the new zone is modified by one of the factors 0;50 and 1;12. It is significant that these two factors—5/6 and 6/5 in fractional form—are precisely the ratios 30:36 and 36:30. I shall give one example from the text:

In Column IV, line 5 we have	♊	2;6.
This is in the fast arc, so we add		36
and obtain	♊	38;6
which exceeds	♊	25
by		13;6.
This multiplied by 0;50 is		10;55
which added to ♊ 25 gives	♊	35;55

or Cancer 5;55, the longitude in line 6.

With these simple rules we may now continue the text as long as we please: First we compute the column of longitudes and form their differences, $\Delta\lambda$. From these we obtain $\Delta\tau$ from

$$\Delta\tau = \Delta\lambda + 12;5,10$$

and so the date columns.

In column VI I have translated the dates in columns I and III into Julian dates; in column VII I have given the longitude of Jupiter corresponding to these dates according to modern tables; and in column VIII are the Babylonian dates of Jupiter's first stationary point, again from modern tables.

When we compare these data it is well to keep in mind that the vernal equinox, from which we count our modern longitudes, played but a secondary role in Babylonian astronomy; the Babylonian zodiac was sidereally fixed, and at the time of the text Babylonian longitudes were about 5° larger than their modern counterparts. Comparing columns IV and VII, the former rounded to nearest integer,

$$3° \leq \lambda(\text{text}) - \lambda(\text{mod.}) \leq 7°,$$

or, allowing for the systematic difference of 5°, that Babylonian and modern longitudes never differ by more than 2°, and rarely by that much. (If instead of the text's scheme we were to use the mean value $\overline{\Delta\lambda} = 33;10°$, the length of this interval would increase from 4° to 11°.)

The Babylonian dates in columns I and III always precede the correct dates in column VIII, but never by more than a week. Here we may recall Ptolemy's complaint in *Almagest* IX 2 ("On the difficulties of constructing a planetary theory") that it is impossible to decide with any accuracy not where, but *when*, a planet becomes stationary. The Babylonian tendency was, it seems, to deem it stationary early.

The quality of the text remains the same to its end (and for a very long while beyond). This seems curious when we consider the crudity of the schemes, and hence wants explanation.

System A: Period Relations

An iterative scheme like the one for finding the longitudes in our text, where any one of the entries determines its successors, is particularly vulnerable to corruption and decay due to computational errors and to imperfections in the scheme itself.

The Babylonian astronomers guarded against mistakes in their calculations by devising various checking rules that allowed them to proceed in steps of many lines (an example in our text is that an advance of 11 lines leads to an effective progress of 5° on the fast arc, 4;10° on the slow).

As to the second point: To prevent the inaccuracies inherent in all approximations to natural phenomena from accumulating arbitrarily, the Babylonians made their schemes obey certain relations that ensured precise return in longitude after astronomically significant periods. Our text is built on the relation that 391 synodic periods of Jupiter correspond to 36 revolutions of Jupiter in the ecliptic and $391 + 36 = 427$ years. Indeed, if we begin with ⥉8;6 in line 1 of our text and calculate its 391 successors correctly, we end with ⥉8;6, and the longitudes have skipped 36 times around the ecliptic.

To convince ourselves of this, we observe that the sequence of longitudes in column IV can be considered as positions of a particle, one time unit apart, if it travels with a speed of 36° per time unit on the fast arc and 30° on the slow. This interpretation, though wrong and anachronistic, leads to the very rules for calculating the longitudes and allows us to ask how long it takes the particle to complete precisely one revolution in the ecliptic. Since the fast arc measures 205°, and the slow 155°, this time interval will be

$$P = \frac{205}{36} + \frac{155}{30} = \frac{391}{36} = 10;51,40.$$

Thus, one revolution lasts nearly 11 time units. Now, in the terms of the text, one "time unit" is one "synodic phenomenon," and a fractional number of synodic phenomena makes little sense. Thus, we put the previous relation in integral form:

391 synodic phenomena ≈ 36 revolutions.

Since during one synodic period the Sun travels one complete revolution in addition to, on the average, Jupiter's synodic arc, the corresponding time interval is $391 + 36 = 427$ years.

System A: General Theory

We have seen one example of what is, in fact, a general theory underlying all planetary arithmetical models of System A. Such a model enables us to calculate corresponding dates and longitudes for consecutive planetary phases of the same kind, for instance, first stationary points Φ. The scheme for finding longitudes is basic: It consists of a generating function and a rule for deriving the synodic arc, $\Delta\lambda$, from it. The generating function, to use an anachronistic term, is a piecewise constant function of longitude—a step function. The ecliptic is divided into arcs $\alpha_1, \alpha_2, \ldots, \alpha_n$, where α_i stands for both an arc and its length; to these are assigned the functional values $w_1, w_2, \ldots, w_n$. We have that

$$\alpha_1 + \alpha_2 + \cdots + \alpha_n = 360°.$$

Provided with an initial value λ_1, we first seek the arc α_i to which λ_1 belongs. If $\lambda_1 + w_i$ still lies in α_i, then $\Delta\lambda$ is w_i, and the longitude of the next Φ is $\lambda_2 = \lambda_1 + w_i$. If not, the interpolation rule comes into play. It is, as given in the procedure texts, that the part of w_i reaching into α_{i+1} is modified in the ratio $w_{i+1} : w_i$, and so on, if more zones are involved.

Thus, a sequence of longitudes may be computed from an initial value. The corresponding sequence of dates runs parallel to it, as it were, for the synodic times are simply

$$\Delta t = \Delta \lambda + C,$$

where C is an appropriate constant.

The principal reason such schemes work so well is that built into them are certain period relations. If we form

$$P = \frac{\alpha_1}{w_1} + \frac{\alpha_2}{w_1} + \cdots + \frac{\alpha_n}{w_n}$$

and write P as an irreducible fraction

$$P = \frac{\Pi}{Z}, \qquad (\Pi, Z) = 1,$$

we can show as before that the planet will return precisely to its initial longitude λ_1 after Π applications of the synodic arc, and not before, and that the phase has traveled Z revolutions in the ecliptic in the process. The integers Π and Z are in all cases astronomically significant parameters. The corresponding total time is Z years for Mercury, $\Pi + Z$ years for Venus, Jupiter, and Saturn, and $2\Pi + Z$ years for Mars.

The Π computed positions arrange themselves very neatly in the ecliptic: They form a set of points that are evenly distributed within each zone; indeed, within the arc α_i any two neighboring possible positions are separated by the interval

$$I_i = \frac{w_i}{Z},$$

as can be shown.

A sequence of positions of consecutive phases can be found easily on the basis of such a distribution of Π intervals in the ecliptic: We let the phase start at any one of the end points and have it progress in steps consisting always of Z consecutive small intervals, regardless of length.

In the scheme of our text we have $\Pi = 391$ and $Z = 36$, so all possible positions are either

$$I_1 = \frac{36°}{36} = 1° \text{ (fast arc)} \quad \text{or} \quad I_2 = \frac{30°}{36} = 0;50° \text{ (slow arc)}$$

apart (there may be, in fact there are in our text, intermediate intervals around the boundaries of the zones). Thus, the fast arc

contains 205 intervals, and the slow 186; in all, there are 391 intervals on the ecliptic. Advancing one line in the longitude column means advancing 36 consecutive intervals in the distribution. Advancing 11 lines means an advance of

$$11 \cdot 36 = 396 = 391 + 5 \equiv 5 \text{ intervals (mod. 391).}$$

The effective advance of 5 intervals after 11 lines means 5° on the fast arc, 4;10° on the slow, the checking rule I cited above.

There remains to consider the scheme for finding dates. The rule

$$\Delta t = \Delta \lambda + C$$

can be derived from the assumption that a certain phase always takes place at a fixed elongation from the sun, but it may also be based on experience. However, once established, whether on theoretical or empirical grounds, there can be no doubt about the value of the constant C: It must be the difference between the mean synodic time, $\overline{\Delta t}$, and the mean synodic arc, $\overline{\Delta \lambda}$. Thus,

---

FIGURE 2. Fig. 2 shows the obverse of the cuneiform text B.M. 36300 from the British Museum's vast collection of Mesopotamian clay tablets. Its columns, read from top to bottom and in order, constitute one long sequence of longitudes, but not dates, of Saturn at consecutive synodic phenomena Γ, Φ, Ψ, Ω, Γ, Φ, Ψ, Ω, etc. (i e., first visibility, first and second stationary points, disappearance), each computed with a System A model for Saturn:

Leo 10° to Pisces 0°: $w = 11;43,7,30°$,

Pisces 0° to Leo 10°: $W = 14;3,45°$

and the period 256 phenomena correspond to 9 revolutions in the ecliptic and (256 + 9 =) 265 years.

The reverse is a continuation of the obverse, and together they are but four columns short of covering a complete period of 265 phenomena of each kind, the longitudes distributing themselves very densely in the ecliptic.

Such a text may have served as a template of sorts: If you are given an initial longitude of a phenomenon, you are sure to find a close approximation to it in the text, and you can simply read off the longitudes of the subsequent phenomena.

This text and others like it are published in A. Aaboe and A Sachs, "Some Dateless Computed Lists of Longitudes of Characteristic Planetary Phenomena from the Late-Babylonian Period." *Journal of Cuneiform Studies*, Vol. XX, pp. 1–33.

1. Babylonian Arithmetical Astronomy

TABLE 3

			Π	Z	i	w_i	$I_i = \frac{w_i}{Z}$
☉	A		46,23	3,45	1	30°	2/15 = 0;8°
					2	28;7,30	1/8 = 0;7,30
	A'		12,23	1,0	1	30	1/2 = 0;30
					2	28;20	17/36 = 0;28,20
☿	A_1	Γ	44,33	14,8	1	1,46	1/8 = 0;7,30
					2	2,21;20	1/6 = 0;10
					3	1,34;13,20	1/9 = 0;6,40
		Ξ	25,13	8,0	1	2,40	1/3 = 0;20
					2	1,46;40	2/9 = 0;13,20
					3	1,36	1/5 = 0;12
	A_2	Σ	20,23	6,28	1	1,47;46,40	5/18 = 0;16,40
					2	2,9;20	1/3 = 0;20
					3	1,37	1/4 = 0;15
					4	2;9;20	1/3 = 0;20
		Ω	11,24	3,37	1	1,48;30	1/2 = 0;30
					2	2,0;33,20	5/9 = 0;33,20
					3	1,48;30	1/2 = 0;30
					4	2,15;37,30	5/8 = 0;37,30
♂	A		2,13	18	1	45	2 1/2 = 2;30
					2	30	1 2/3 = 1;40
					3	40	2 2/9 = 2;13,20
					4	1,0	3 1/3 = 3;20
					5	1,30	5 = 5
					6	1,7;30	3 3/4 = 3;45
♃	A		6,31	36	1	36	1 = 1
					2	30	5/6 = 0;50
	A_1		16,19	1,30	1	36	2/5 = 0;24
					2	30	1/3 = 0;20
	A'		6,31	36	1	30	5/6 = 0;50
					2	33;45	15/16 = 0;56,15
					3	36	1 = 1
					4	33;45	15/16 = 0;56,15
	A''		6,31	36	1	30	5/6 = 0;50
					2	33;45	15/16 = 0;56,15
					3	36	1 = 1
					4	33;45	15/16 = 0;56,15
	A'''		4,53	27	1	30	1 1/9 = 1;6,40
					2	33;45	1 1/4 = 1;15
					3	36	1 1/3 = 1;20
					4	33;45	1 1/4 = 1;15
♄	A		4,16	9	1	11;43,7,30	1 29/96 = 1;18,7,30
					2	14;3,45	1 9/16 = 1;33,45

TABLE 3 (continued)

α_i	$v_i = \dfrac{\alpha_i}{I_i}$	$P = \dfrac{\Pi}{Z}$	α_i
3,14°	24,15	12;22,8	♍ 13 – ♓ 27
2,46	22,8		♓ 27 – ♍ 13
2,44;30	5,29	12;23	♍ 29 – ♓ 13;30
3,15;30	6,54		♓ 13;30 – ♍ 29
2,45	22,0	3;9,7,38,...	♌ 1 – ♉ 16
2,14	13,24		♉ 16 – ♈ 30
1,1	9,9		♊ 0 – ♌ 1
1,50	5,30	3;9,7,30	♋ 6 – ♎ 26
2,14	10,3		♎ 26 – ♓ 10
1,56	9,40		♓ 10 – ♋ 6
1,30	5,24	3;9,7,25,..	♋ 0 – ♍ 30
1,36	4,48		♎ 0 – ♉ 6
1,29	5,56		♉ 6 – ♈ 5
1,25	4,15		♈ 5 – ♊ 30
3,0	6,0	3;9,7,27,..	♋ 0 – ♐ 30
1,0	1,48		♉ 0 – ♒ 30
1,0	2,0		♓ 0 – ♈ 30
1,0	1,36		♉ 30 – ♊ 30
1,0	24	7;23,20	♉ ♊
1,0	36		♋ ♌
1,0	27		♍ ♎
1,0	18		♏ ♐
1,0	12		♑ ♒
1,0	16		♓ ♈
3,25	3,25	10;51,40	♐ 0 – ♊ 25
2,35	3,6		♊ 25 – ♐ 30
3,22	8,25	10;52,40	♐ 5 – ♊ 27
2,38	7,54		♊ 27 – ♐ 5
2,0	2,24	10;51,40	♋ 9 – ♍ 9
53	56;32		♍ 9 – ♉ 2
2,15	2,15		♉ 2 – ♉ 17
52	55;28		♉ 17 – ♋ 9
2,0	2,24	10;51,40	♋ 5;45 – ♍ 5;45
56;15	1,0		♍ 5;45 – ♉ 2
2,15	2,15		♉ 2 – ♉ 17
48;45	52		♉ 17 – ♋ 5;45
2,0	1,48	10;51,6,40	♋ 9 – ♍ 9
48	38;24		♍ 9 – ♐ 27
2,20	1,45		♐ 27 – ♉ 17
52	41;36		♉ 17 – ♋ 9
3,20	2,33;36	28;26,40	♌ 10 – ♒ 30
2,40	1,42,24		♓ 0 – ♌ 10

the constant is readily derived from Π and Z. Here we need that

$$1 \text{ year} = 12;22,8 \text{ synodic months} = 12;22,8 \cdot 30 \text{ tithis} = 6,11;4^\tau,$$

a basic parameter of Babylonian theoretical astronomy (it is, e.g., the period P of the longitude columns in lunar ephemerides of System A). With these parameters we obtain

$$\overline{\Delta\lambda} = \frac{Z \cdot 6,0°}{\Pi}$$

and, for Jupiter and Saturn,

$$\overline{\Delta t} = \frac{\Pi + Z}{\Pi} \cdot 6,11;4^\tau$$

(for Mars, $\Pi + Z$ should be replaced by $2\Pi + Z$).

Thus,

$$C = \overline{\Delta t} - \overline{\Delta\lambda} = 6,11;4 + \frac{Z}{\Pi} \cdot 11;4$$

$$= 12^m + \left(1 + \frac{Z}{\Pi}\right) \cdot 11;4^\tau$$

(the rule's direct translation of degrees into *tithis* reflects that the sun travels very nearly one degree per *tithi*). With $\Pi = 391$ and $Z = 36$, we find

$$C = 12^m + 12;5,8,7,\ldots^\tau,$$

which compares well with our text's use of 12;5,10 for the last part.

In Table 3 I gathered the parameters of all planetary models of System A and included those of the monthly motion of the sun from the lunar theory. In addition, N. T. Hamilton recovered parts of a System A scheme for Venus from *ACT* No. 1050, recently published, but its parameters are not included in Table 3.

Planetary Theory: System B

Table 4 gives the first 22 of at least 62 lines of *ACT* No. 620, an ephemeris for Jupiter at opposition for at least the years S.E. 127 to 194. It is arranged much like the previous text except that here both difference columns are included, and we shall first examine column IV, $\Delta\lambda$.

TABLE 4

No. 620	I	II	III	IV	V
l.	year	Δτ	month & date	Δλ	λ
0	2,7	49;42	V 27;36	37;37	♓ 24;31
1	2,8	47;54	VII 15;30	35;49	♈ 30;20
2	2,9*	46;6	IX 1;36	34;1	♊ 4;21
3	2,10	44;18	IX 15;54	32;13	♋ 6;34
4	2,11	42;30	X 28;24	30;25	♌ 6;59
5	2,12**	40;42	XI 9;6	28;37	♍ 5;36
6	2,13	41;47,30	XII 20;53,30	29;42	♎ 5;18
7	2,15	43;35,30	I 4;29	31;30	♏ 6;48
8	2,16	45;23,30	II 19;52,30	33;18	♐ 10;6
9	2,17*	47;11,30	IV 7;4	35;6	♑ 15;12
10	2,18	48;59,30	IV 26; 3,30	36;54	♒ 22;6
11	2,19	49;27	VI 15;30,30	37;22	♓ 29;28
12	2,20*	47;39	VIII 3; 9,30	35;34	♉ 5;2
13	2,21	45;51	VIII 19; 0,30	33;46	♊ 8;48
14	2,22*	44;3	X 3; 3,30	31;58	♋ 10;46
15	2,23	42;15	X 15;18,30	30;10	♌ 10;56
16	2,24	40;27	XI 25;45,30	28;22	♍ 9;18
17	2,25*	42; 2,30	XII₂ 7;48	29;57	♎ 9;15
18	2,27	43;50,30	I 21;38,30	31;45	♏ 11
19	2,28*	45;38,30	III 7;17	33;33	♐ 14;33
20	2,29	47;26,30	III 24;43,30	35;21	♑ 19;54
21	2,30	49;14,30	V 13;58	37;9	♒ 27;3
22	2,31**	49;12	VI₂ 3;10	37;7	♈ 4;10
.	.	.	.	.	.
.	.	.	.	.	.
.	.	.	.	.	.

The entries in column IV decrease regularly by the amount $d = 1;48°$ per line until a minimum is passed between lines 5 and 6. From line 6 the values increase, again by $1;48°$, until a maximum is passed between lines 10 and 11, at which point they begin to decrease, and so on. If we plot $\Delta\lambda$ as a function of line number, we get a piecewise linear graph like the one in Figure 3. One gets from an ascending branch to a descending branch of such a zigzag function, as we call it, by following a simple reflection rule (often stated in procedure texts); if the application of the line-by-line difference d (here $1;48°$) leads to a value larger than a certain fixed maximum M (here $38;2°$), then the excess over M is subtracted from M to yield the next value of the function, and symmetrically about the minimum m (here $28;15,30°$). The reflection in the maximum between lines 10 and 11 is then

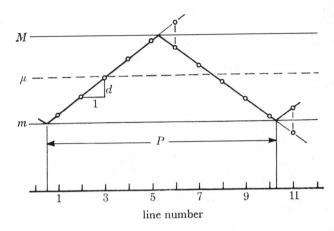

FIGURE 3.

executed thus:

$$\begin{array}{rr}\text{Col. IV, l. 10:} & 36;54°\\ +d: & 1;48°\\ \hline & 38;42°\\ -M: & 38;\ 2°\\ \hline & 0;40°\end{array}$$

which subtracted from M gives: $37;22°$,

the entry in line 11.

The zigzag function, sometimes refined in various ingenious ways, is one of the two basic modes of describing a simply periodic component of a more complex astronomical phenomenon—the other is the step function of System A. It is easy to compute, and it has a simply controlled period:

$$P = \frac{2(M-m)}{d}.$$

With the parameters of the present text, $M = 38;2°$ and $m = 28;15,30°$, $d = 1;48°$, we obtain

$$P = 10;51,40 = \frac{391}{36}.$$

We recognize this as precisely the period of the System A scheme for Jupiter. There it implied that $\Pi = 391$ applications of the synodic arc lead to precisely $Z = 36$ revolutions in the ecliptic. Here the period relation is that Π lines lead to precise return in $\Delta\lambda$ and embrace Z "waves" of the zigzag function, but it says nothing about the longitudes themselves. To ensure return in longitudes, we should require that the mean synodic arc be

$$\overline{\Delta\lambda} = \frac{Z \cdot 360°}{\Pi} = \frac{36 \cdot 360}{391} = 33;8,44,48,\ldots°,$$

and our zigzag function obliges by having as its mean value

$$\mu = \tfrac{1}{2}(M + m) = 33;8,45°.$$

In column II of No. 620, $\Delta\tau$ is represented as a zigzag function of parameters

$$M = 50;7,15^\tau, \qquad m = 40;20,45^\tau, \qquad d = 1;48^\tau,$$
$$\mu = \tfrac{1}{2}(M + m) = 45;14^\tau.$$

In this text we do not quite have that the difference between $\Delta\tau$ and $\Delta\lambda$ is constant, but the amplitudes and differences, and so the periods, of the two zigzag functions are identical. Thus, the two functions are slightly out of phase, for no reason that I can see other than numerical convenience. However, the difference between the mean values of columns II and IV is 12;5,15, which is close to the corresponding value 12;5,10 of the previous text.

Daily Planetary Positions

We have now seen how the date and longitude of the planetary phases were computed. The day-by-day progress of a planet was calculated by interpolation between neighboring phases. For Jupiter (which here, as in other respects, is best documented among the planets) we find several such interpolation schemes of varying degree of complexity—the procedure text *ACT* No. 810 alone contains three variants.

Table 5 presents the most elaborate and sophisticated example of such schemes. It is the result of P. Huber's ingenious reconstruction and linking of two *ACT* texts, and I have excerpted the most interesting parts. The first column gives year and date (years in S.E., dates in days, not *tithis*), the last Jupiter's longitudes, and the first and second columns the second and first differences of the longitudes. The phases divide the text into stretches of constant third difference, as indicated at the brackets to the table's right.

It is worthy of note that in 164 B.C. ($= -163$), the year of the text, Jupiter's retrograde arc was 9.8° (Tuckerman). This is a full

TABLE 5

	T		$\Delta^2\lambda$	$\Delta\lambda$	λ	
1.	147 IX	1	0	+12,40	29 ♏	Γ
		2	− 6	+12,39,54	29;12,39,54	
		3	− 12	+12,39,42	29;25,19,36	
		4	− 18	+12,39,24	29,37,59	
5.		5	− 24	+12,39	29,50,38	
		6	− 30	+12,38,30	3,16,30 ♐	
		7	− 36	+12,37,54	15,54,24	
		8	− 42	+12,37,12	28,31,36	
		9	− 48	+12,36,24	41, 8	
10.		10	− 54	+12,35,30	53,43,30	$\Delta^3\lambda = -0;0,0,6°$
		11	−1, 0	+12,34,30	1; 6,18	
		⋮	⋮	⋮	⋮	
119.	148 I	1	−11,48	+ 57,54	16; 6,36 ♐	
		2	−11,54	+ 46	16; 7,22	
		3	−12	+ 34	16; 7,56	
		4	−12, 6	+ 21,54	16; 8,17,54	Φ
		5	−12,12	+ 9,42	16; 8,27,36	
		6	−12, 2	− 21,44	16; 8, 5,52	
125.		7	−11,52	− 33,36	16; 7,32,16	
		8	−11,42	− 45,18	16; 6,46,58	
		9	−11,32	− 56,50	16; 5,50, 8	
		10	−11,22	− 1, 8,12	16; 4,41,56	$\Delta^3\lambda = +0;0,0,10°$
		⋮	⋮	⋮	⋮	
181.	III	4	−2,32	−7,12, 8	11;46,14, 8 ♐	
		5	−2,22	−7,14,30	11;38,59,38	Θ
		6	+2,32	−7,11,58	11;31,47,40	
		7	+2,42	−7, 9,16	11;24,38,24	
		⋮	⋮	⋮	⋮	$\Delta^3\lambda = +0;0,0,10°$
238.	V	1	+11,42	− 35,58	7;22, 7,30 ♐	
		2	+11,52	− 24, 6	7;21,43,24	
240.		3	+12, 2	− 12, 4	7;21,31,20	
		4	+12,12	+ 24,16	7;21,55,36	Ψ
		5	+12, 6,30	+ 36,22,30	7;22,31,58,30	
		6	+12, 1	+ 48,23,30	7;23,20,22	
		⋮	⋮	⋮	⋮	$\Delta^3\lambda = -0;0,0,5,30°$
374.		·	+0, 0,30	+13,50, 1,30	28;14,59,23,30	

ACT Nos. 654 & 655

degree more than what we find in the text. The motivation of this fine scheme cannot, then, be found among the concerns of practical astronomy, for if the Babylonian astronomers could tolerate an error of one degree in the length of Jupiter's retrograde arc, they surely had no need for so elaborate a scheme to account satisfactorily for Jupiter's daily motion.

Lunar Theory

Babylonian lunar theories are classified as belonging to System A or System B according to their manner of accounting for monthly solar (and lunar) progress: System A employs a step function of longitude; System B a zigzag function of line number. However, this is not the only, nor even the most significant, difference between the two systems.

The relations between Systems A and B for the moon are not at all clear. The two systems were in use at the same time, in both Babylon and Uruk, so one did not replace the other. An analysis of System A reveals a tight and consistent inner structure, deducible from a rather small number of reasonable basic assumptions—most of them are not mentioned in the procedure text. In comparison, System B seems a collection of parts that are at best but loosely connected and in one instance (column J) of quite unwarranted complexity.

Lunar Theory: System A

In Table 6 I reproduced our reconstruction of a small fragment of a lunar ephemeris of System A (*ACT* No. 20): The preserved text is inside the dotted lines. I chose this example rather than a better-preserved text to emphasize that all lunar ephemerides of System A are strictly connectible (except for the last visibility columns) and thus are slices, as it were, of one great continuing ephemeris that we have been able to bring under control and reconstruct, particularly with the aid of electronic computers. This feature seems quite in harmony with the inner consistency of System A, and we find nothing like it in the texts of System B.

Column T gives year (S.E.) and month, and everything in a horizontal line concerns the situation near a syzygy in that month—on the obverse the conjunction at the end of the month, on the reverse the opposition at month's middle.

1. Babylonian Arithmetical Astronomy

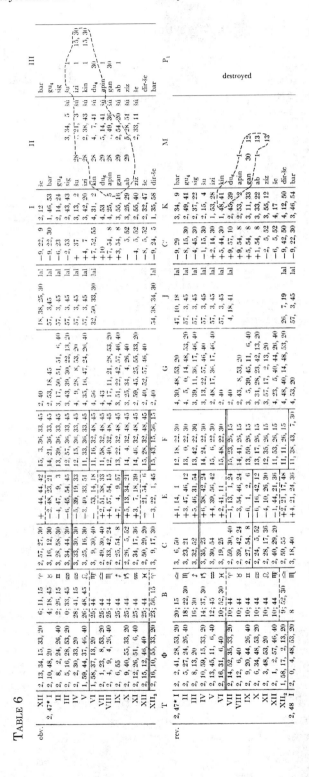

Table 6

Column Φ is a zigzag function whose period is the anomalistic month; in fact, it is in phase with the apparent lunar velocity (column F). It serves as a means of injecting lunar anomaly into the theory. We have recently found that column Φ denotes the length of 223 successive lunations, beginning at the syzygy for which Φ is listed, minus 6585 days and assuming a constant monthly solar progress of 30°, but it had long been known how Φ served as the basis of finding the later column G. I shall not explain here the very sophisticated methodology underlying the relation of Φ to G (and to other similar functions).

In column B we find the longitude of the moon at syzygy or, more correctly, monthly positions of the sun on the obverse, and monthly positions of the sun increased by 180° on the reverse. The sun moves according to the System A scheme whose parameters are given in Table 3 and whose period $P = 12;22,8$ months is the length of year that pervades all of Babylonian astronomy (it serves as the value of the sidereal, tropical, and anomalistic year; the Babylonian astronomers were unaware that these three years were of different lengths). It should be noted that lunar anomaly has so small an effect on the *position* of syzygies that it may safely be ignored. Column C gives the length of daylight, and column E is lunar latitude (in units, "barleycorns," 72 of which equal one degree; the integral barleycorns occupy the first two sexagesimal places). Lunar latitude turns out to be a very simple function—a slightly modified zigzag function—of the moon's elongation from the ascending node; this nodal elongation is easily found from column B giving lunar longitude combined with the underlying assumption that the node retrogresses by the constant amount 1;33,55,30° per month, and some initial position of the node. Eclipse warnings would be issued whenever the new or full moon has smallest latitude at a nodal crossing; this is done with an eclipse magnitude function depending simply on column E, but which I have not bothered to reconstruct.

Columns G and J together give the excess over 29 days of the time from conjunction to conjunction, or opposition to opposition, where G depends on lunar, and J on solar anomaly. The eclipse magnitude Ψ predicts lunar eclipses very well, but for solar eclipses—treated precisely the same as lunar eclipses—it can serve only as a possible maximum. Column C' gives a correction due to the variation in length of daylight, column K the sum of G, J, and C', and column M lists the moment of syzygy (on

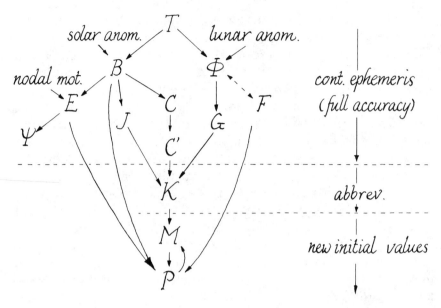

FIGURE 4. Lunar system A

the obverse date and time degrees of conjunction *before* sunset of the day), and the moments proceed from line to line by the amount $-K$ on the obverse, and K on the reverse. Finally, column P, preserved on obverse only, gives information about the visibility of the new crescent; "1" means that the previous month turned out to be full (30 days long), "30" means that it was hollow (29 days long), and the following numbers give the computed time from sunset to moonset.

In Figure 4 I show in schematic form the relations between the columns of lunar System A.

Lunar Theory: System B

Figure 5 is a photograph of the reverse of *ACT* No. 122 (BM 34580 + ···), one of the best-preserved lunar ephemerides. It was written in S.E. 209 IX 18, as the scribe tells us in the colophon, and it treats new moons for the years S.E. 208, 209, and 210.

I cannot give a transcription of the text here, but I shall mention a few of the characteristic features of lunar System B, to which this text belongs. Monthly progress of the sun and moon

FIGURE 5. Lunar ephemeris computed according to System B for consecutive new moons for the years S.E. 208, 209, and 210, and written in Uruk S.E. 209, month IX, day 18 (=103 BC, Dec. 22/3). [ACT No. 122 = BM 34580 + ⋯.]

is computed according to a zigzag function of parameters

$$M = 30;1,59°, \qquad d = 0;18,$$
$$m = 28;10,39,4°, \qquad \mu = 29;6,19,20,$$

and period $P = 12;22,8,53,20$ and with

$$\frac{360}{\mu} = 12;22,7,51,\ldots,$$

where both last numbers are close to the canonical year length of 12;22,8 months.

The remaining columns have functions analogous to those of System A, but most of them are computed in isolation from the rest. I shall only draw attention to column G, the eighth column of a standard ephemeris. Here, as in System A, the time from one conjunction to the next is $29^d + G + J$, where G's period is the anomalistic month, and J's the year. Column G is a zigzag function of mean value

$$\mu = 3,11;0,50° = 0;31,50,8,20 \text{ day},$$

to convert time degrees into days ($1^d = 6,0°$). Since column J has $0°$ as its mean value, we then have the following Babylonian value for the mean synodic month:

$$29 \text{ days} + \mu = 29;31,50,8,20 \text{ days}.$$

I draw attention to this parameter, for it is *precisely* the value adopted by Hipparchus and, following him, by Ptolemy for the length of the mean synodic month (*Almagest* IV, 2), as already Kugler observed. This is but one example of the pervasive influence of the Babylonians on Hellenistic Greek astronomy, and one of many supports for my conviction that all subsequent activity in scientific astronomy—if not in the exact sciences—is descended in direct line from the creation of the anonymous Babylonian astronomers.

Survival of Babylonian Methods

Had I submitted my manuscript a decade earlier, I would have ended the chapter on Babylonian astronomy here, but in the course of that interval startling new evidence about the transmission of Babylonian astronomical knowledge to Hellenistic Greece, or rather Egypt, has come to light.

In the works of Hipparchus and Ptolemy we recognize, as said, many parameters of Babylonian origin, and Ptolemy quotes Babylonian observations. However, we find no evidence whatsoever that they knew anything about the Babylonian arithmetical methods. I myself believed, but it was no more than a belief, that the mere existence of Babylonian ephemerides had convinced Hipparchus that a quantitative, mathematical description of certain astronomical phenomena, capable of yielding numerical predictions that could be tested against observations, was both possible and desirable. Consequently, I thought, he adapted Greek qualitative, geometrical models to quantitative ends, inventing trigonometry for the purpose. For this, of course, he did not have to know how these ephemerides were computed, and whether he, or Ptolemy, had such knowledge, but chose not to write about it in their surviving works, remained an unanswerable question.

So it still does, but others in the Hellenistic world certainly had control of Babylonian methods, as we have now learned. In 1988 Neugebauer published a small fragment of a Greek papyrus, a photograph of which had been sent him (see Figure 6). It contains a column of numbers that, Heaven help us, forms 32 consecutive lines of a column G of Babylonian lunar theory, System B, the very column that, as I said earlier, has a mean value in which Kugler recognized Hippachus's and Ptolemy's length of the mean synodic month. Such a column makes no sense in isolation, so it is a safe inference that the techniques of one of the systems of Babylonian lunar theory were known in Roman Egypt in the first half of the first century A.D. (the papyrus is dated by the handwriting).

This was, indeed, an astonishing discovery, but it was only the tip of the iceberg, for in the 1990s Alexander Jones, the brilliant young Canadian scholar, edited and analyzed a great number of astronomical papyri in a series of papers, culminating in his impressive *Astronomical Papyri from Oxyrhynchus* in a massive volume in 1999.

These papyri from Roman Egypt, dating from the first three or four centuries A.D., contain many wonderful and unexpected things. Of interest here is that we find ample evidence for Babylonian schemes, not only for the moon, but for the planets as well, competently adapted to the local calendar, when required.

64   1. Babylonian Arithmetical Astronomy

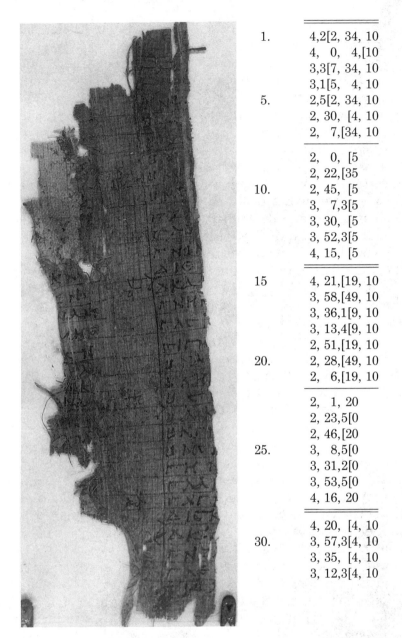

1.	4,2[2, 34, 10
	4, 0, 4,[10
	3,3[7, 34, 10
	3,1[5, 4, 10
5.	2,5[2, 34, 10
	2, 30, [4, 10
	2, 7,[34, 10
	2, 0, [5
	2, 22,[35
10.	2, 45, [5
	3, 7,3[5
	3, 30, [5
	3, 52,3[5
	4, 15, [5
15	4, 21,[19, 10
	3, 58,[49, 10
	3, 36,1[9, 10
	3, 13,4[9, 10
	2, 51,[19, 10
20.	2, 28,[49, 10
	2, 6,[19, 10
	2, 1, 20
	2, 23,5[0
	2, 46,[20
25.	3, 8,5[0
	3, 31,2[0
	3, 53,5[0
	4, 16, 20
	4, 20, [4, 10
30.	3, 57,3[4, 10
	3, 35, [4, 10
	3, 12,3[4, 10

FIGURE 6. The papyrus (P. Colker) that offered our first evidence of knowledge in Hellenism of the technical details of Babylonian arithmetical astronomy. Its provenance is not known, but it is written in a hand of the first half of the first century A.D.

It contains remnants of two columns, the first of which has so far defied restoration. The second column, to the right of the vertical ruling, presents 32 lines of a column G of Babylonian lunar System B, as described at the end of this chapter. A restored transcription is appended.

Thus, we have learned that Babylonian arithmetical methods survived well into Roman times, very likely kept alive by the needs of astrologers, but without leaving any trace in the astronomy of the Ptolemaic tradition. Here we may recall the opening lines of Ode I, xi, by Horace (65–8 B.C.), the lovely ode that contains the admonishment *carpe diem*:

Tu ne quaesieris—scire nefas—quem mihi, quem tibi,
finem di dederint, Leuconoë, nec Babylonios
temptaris numeros ...

[Do not ask, Leuconoë (it is impossible (impious) to know), what end the gods have ordained for me, for you, nor put the Babylonian numbers (tables) to the test ...]

2
Greek Geometrical Planetary Models

Greek Geometrical Models

Higher Greek mathematics is mostly concerned with geometry, so it is not too surprising that the Greek detailed planetary models were geometrical. The aim of such models was at first to mimic the behavior of a planet, which, in the case of Saturn, is indicated in Figure 1, with latitude exaggerated four times.

The figure does not show that the Sun plays a particular role for the planet's behavior: When Saturn is at the middle of its retrograde arc, it is also in opposition, Θ, to the Sun, that is, its longitude differs from the Sun's by 180°. The same holds for Jupiter and Mars, the other two outer planets. However, an inner planet is at the middle of its retrograde arc when it is in what we call "inferior conjunction" to the sun. I should here remark that the ancients divided the planets into two classes identical with our inner and outer planets, but according to the criterion of whether they can reach opposition (our outer planets) or not (our inner planets, which are, as it were, tethered to the sun on a short leash). Please note that I am not attempting to explain these things. I merely state them as facts discoverable by fairly short runs of sustained observations, as are the following crude period relations.

Let us now examine Saturn's behavior a bit more closely (see again Figure 1). It is a matter of simple counting and record-keeping that 29 loops lead to just about one revolution in the ecliptic in a time interval of some 30 years.

For the other planets similar relations hold, but with different numbers: For Jupiter we find that 11 loops bring the planet nearly once around the ecliptic in the course of some 12 years.

For Mars the situation is a little more complicated. Here the

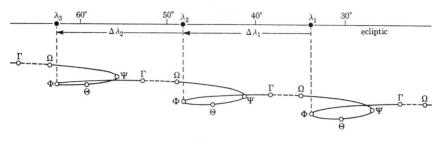

FIGURE 1.

shortest time interval that leads to a near return of the loops in the ecliptic is 32 years. In the course of this time span, a synodic phenomenon, say, first stationary point Φ, skips around the ecliptic twice, very nearly, taking with it 15 loops. However, Mars itself takes an extra turn around the ecliptic when it travels from one Φ to the next, so the planet will have performed 15 (the number of loops) plus 2 (the number of the phenomena's revolutions), or 17, revolutions in the ecliptic in 32 years.

For Venus it happens that a very simple period relation is also very good, and it has been independently recognized in several ancient cultures. It states that in eight years Venus goes through its synodic cycle five times and travels very nearly eight times around the ecliptic.

For Mercury, a planet we shall ignore here, the crudest estimate of these parameters is that three synodic periods correspond very roughly to one revolution in the ecliptic or one year.

To return to the problem of Greek geometrical planetary models. It is perhaps natural to construct such a model of two components: one that makes the planet oscillate back and forth and another that drags the former along the ecliptic. At any rate, that is what was done in the two types of models we shall now consider (the latter comes in two versions).

The Homocentric Spheres of Eudoxos

The first model is the homocentric spheres of Eudoxos, the eminent Greek mathematician and philosopher who lived during the first half of the fourth century B.C. In mathematics he is particularly renowned for his theory of proportions and of exhaustion (integration), both of which we find in Euclid's *Elements*; no work of Eudoxos is preserved in its original form. His major

2. Greek Geometrical Planetary Models

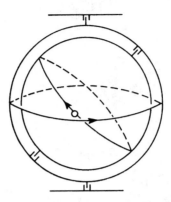

FIGURE 2.

contribution to astronomy was brilliantly pieced together by Schiaparelli from secondary and tertiary sources, mostly Simplicius's commentary to Aristotle's *De caelo* ("On the heavens"). Before we proceed I want to make it clear that "homocentric" means "concentric" or that the spheres have a common center that further is occupied by the observer.

The model's first component, the one that produces an oscillation, consists of two concentric spheres, the inner being able to rotate relative to the outer, and the outer being able to rotate relative to a frame that for the moment we shall consider as fixed (see the very schematic drawing in Figure 2). We are now interested in the behavior of a point (the planet) on the "equator" of the inner sphere, in fact, the very point that is also on the "equator" of the outer sphere when the motion begins.

We now let the inner sphere rotate with a certain angular velocity (say, in degrees or rotations per time unit: day, month, or what have you) relative to the outer sphere, and the outer with the *same* angular velocity in the *opposite* direction but now relative to the frame.

If the two axes of rotations were not tilted against each other, but coincided, the "planet" would not move at all relative to the frame, for the rotation of the outer sphere would cancel that of the inner. Now, if we tilt the two axes a little against each other, it seems reasonable enough that the "planet" will wiggle about its initial position a little bit and with a period equal to the common period of the two rotations. But here the intuition of most people (I am among them) gives out.

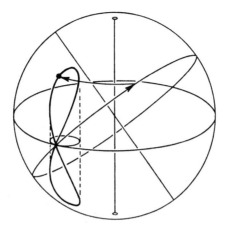

FIGURE 3.

The path of the "planet" for various tilts of the two axes was analyzed by Schiaparelli by means of spherical trigonometry and later by Neugebauer, who used only mathematical techniques available to early Greek geometers. This path turns out to be a curve on the spherical surface that looks much like a figure-eight (see Figure 3)—its ancient name is a *hippopede* (a horse-fetter, i.e., the rope loops used to hobble a horse's forelegs so it cannot stray). In fact, the hippopede turns out to be the intersection of a sphere with a (small) right cylinder, tangent to the sphere at one point as shown in Figure 3, which may have been known in antiquity, for it is shown by ancient means by O. Neugebauer in *Exact Sciences in Antiquity*. Here only one sphere is drawn, but both axes and their "equators" are represented.

We now place this apparatus so that the vertical line of symmetry of the figure-eight is the ecliptic, and the whole thing is given a forward motion in longitude (this may be achieved by yet another sphere rotating, with the ecliptic as its "equator" relative to the fixed stars).

The third motion is direct or forward in longitude (toward the left in Figure 4). The figure shows what the resulting path of the "planet" would look like when observed from the common center of the spheres for different forward motions. (I should warn the reader that the horizontal scales are not to be taken as the ecliptic—the ecliptic is a horizontal line parallel to the scales, but bisecting the now-horizontal figure-eight.) At the top we

70   2. Greek Geometrical Planetary Models

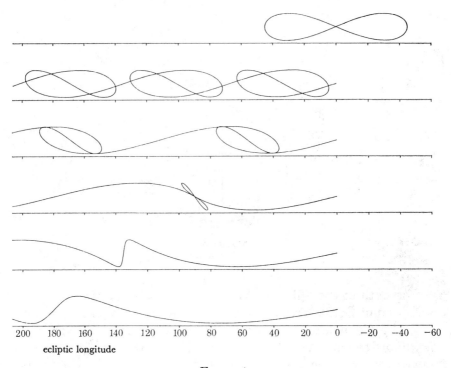

FIGURE 4.

have the hippopede itself, and as the forward motion increases, we first see a curve that indeed looks somewhat like the apparent path of a planet with its stationary points and retrogradations. If, however, the forward motion is too swift, then the "planet" can no longer become retrograde, but is merely slowed down, as shown in the two lowest graphs of Figure 4.

When we toy like this with various velocities, we assume that we can assign them at will. But when we deal with a specific planet, that is not so. Indeed, the period of the motion in the hippopede (i.e., of the rotations of the first two spheres) must be the planet's synodic period, and the forward motion in longitude must be the (mean) synodic arc per synodic period. Both of these are determined within narrow bounds by even crude period relations of the sort I just mentioned. Our only really free choice in this model is, then, of the inclination of the two axes in Figure 2.

Now, of the simple period relations I mentioned earlier, I must emphasize the ones for our neighbors in the solar system, Venus and Mars.

For Venus we have, as we saw,

5 synodic cycles ≈ 8 years

≈ 8 revolutions in the ecliptic of Venus,

and for Mars:

15 synodic cycles ≈ 17 revolutions ≈ 32 years.

In terms of Eudoxos's homocentric spheres, Venus must then travel through its hippopede 5 times in 8 years, and the hippopede must be carried around 8 times in the ecliptic in the same 8 years. Analogously, Mars must travel through its hippopede 15 times in 32 years, while the hippopede is carried 17 times around in the ecliptic in the same 32 years.

It can be shown that in both cases the forward motion is so swift that whatever the inclination of the axes of the two inner spheres we must always end with a situation like that represented in the two lowest graphs in Figure 4: Venus and Mars just cannot become retrograde, but are merely slowed down. This result was already derived by Schiaparelli, but the proof, though elementary, is rather messy, so I shall omit it and treat more carefully the two models I shall consider next, where some of the same issues arise.

Let me, however, already here reveal what it is I am after with my arguments. We have become accustomed to consider the role of a geometrical model of the motion of, say, a planet is that of serving as a basis for computing the planet's position at a certain time in some relevant coordinate system. So it has, in fact, been since, and including, the work of Ptolemy.

Now, deriving a planetary position (say, in ecliptic longitude and latitude) from an Eudoxian model would surely involve spherical trigonometry, a subject that very likely was not addressed seriously before Menelaos (first century A.D.).

For this reason alone—that the required mathematical techniques for extracting numerical coordinates from the Eudoxian models were not at hand until well after the models' invention—we could conclude that the models' original purpose was *qualitative* rather than *quantitative*.

But even if that had not been so, the qualitative character of these models would be amply established by their inability to make Mars and Venus retrograde.

There is a hint, albeit very faint, that someone, probably Eudoxos or his follower Calippus (Kalippos), was aware that the

Martian model fails to produce retrogradations. According to Simplicius, the period of Mars in its hippopede was taken to be 260 days, or precisely one-third of 780 days, which is a good value of Mars's synodic period. Indeed, our crude relation before, that 15 synodic periods correspond to 32 years (i.e., $32 \cdot 365.25 = 11{,}688$ days) yields $11{,}688/15 = 779.2$ days for this parameter. Since Mars in the same 32 years revolves 17 times in the ecliptic, the hippopede must be carried once around the ecliptic in $11{,}688/17 = 687.5$ days.

If we maintain this last carrying motion but adopt 260 days as a synodic period—the period of travel through the hippopede—the new planet can very well become retrograde, but we shall now have three retrogradations when Mars has but one.

The motive for the adoption of the shorter period of the motion in the hippopede may then have been a desire for obtaining retrogradations, but at the cost of getting thrice as many as one should. If this were so, someone must have been aware of the deficiency of a Eudoxian model for Mars.

Epicyclic Models

The next model we shall consider—the epicyclic model— is of unknown origin. We know from Ptolemy that Apollonios of Perga, the great mathematician who wrote on conic sections about 200 B.C., proved an elegant theorem about stationary points in epicyclic models, so the invention of this kind of model antedates Apollonios if it is not, in fact, his own, which I doubt.

Consider now Figure 5. The plane of the paper is the plane of the ecliptic viewed from the north. The observer sits at O on the earth, and the vertical ray from O points to a certain fixed star. The point C travels uniformly, relative to the direction from O to the star, and counterclockwise on a circle—the *deferent*—with O as its center. The planet P in turn travels on a circle—the *epicycle*—with C as its center and a radius CP less than that of the deferent (OC). The planet P's motion on this second circle is also uniform, and we reckon it, as the ancients always did, from the deferent's radius OC or its extension to A; A is called the epicycle's apogee (the point farthest from the earth), and Π its perigee (the point closest to the earth).

We recognize here, once more, the two components of a planetary model, one that makes the planet oscillate around a mean position and the other that carries the first along the ecliptic.

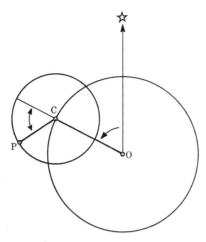

FIGURE 5.

The former component is the epicycle; indeed, if we arrest C and observe P's motion from O, we shall see the epicycle edge on, and P will seem to oscillate in the ecliptic back and forth around C. The period of this oscillation ought then to be the synodic period—for Venus, 5 revolutions in 8 years, and so on.

The latter component consists of the motion of C, and with it the entire epicycle, along the deferent. The rate of this motion must be the mean synodic arc per synodic period—for Venus 8 revolutions in 8 years, for Saturn 1 revolution in some 30 years, and so forth.

Before we proceed I should emphasize that if all we are interested in is the *direction* from the observer O to the planet P, the absolute size of the model is of no concern to us. In fact, Ptolemy shows in *Almagest III*, 3 that the direction OP remains the same if you subject the model to a dilation or contraction (i.e., a similarity transformation) with respect to O.

We shall then follow Ptolemy when we use OC, the radius of the deferent, as our unit for measuring the dimensions of the model—he, however, employs the sexagesimal system and calls the radius of the deferent 60 parts.

Let us now return to Figure 5. As already said, we shall normalize and set $OC = 1$, and the epicyclic radius $CP = r < 1$, measured with OC as unit.

As I indicated in the figure, P can in principle move clockwise or counterclockwise on the epicycle, relative to the deferent's radius OC or its extension to the epicycle's apogee.

If P moves clockwise—we shall call this sense of rotation "wrong," for reasons to be set forth later—the planet ought to become retrograde when it is near the apogee A of its epicycle. If it moves counterclockwise—the "right" way, as we shall see—it becomes retrograde when in the neighborhood of the epicycle's perigee.

We have, in fact, evidence for the use of both senses of rotation on the epicycle. Our main source for the wrong sense is a long astrological Greek papyrus, Papyrus Michigan 149, written in a hand most likely of the second century A.D., that is, about the time of Ptolemy. However, that is not to say that the astronomical parts of the papyrus reflect the high level of Ptolemy's work. On the contrary, P. Mich. 149 preserves, encapsulated in fossil form, as it were, purely astronomical parts that represent a variety of older and more primitive astronomical schemes and theories, even though these different schemes are not at all compatible. This situation is quite common in astrological texts, and that is why they are of great interest, not only to historians of astrology, but to historians of astronomy as well.

In P. Mich. 149 we find a passage describing the behavior of Venus during one synodic cycle in terms of the planet's distance from the Earth. The description is a bit vague, but it leaves no doubt whatsoever that the planet is assumed to move the wrong way around on its epicycle. We would expect a corresponding passage for outer planets, but here the papyrus is destroyed. Even so, it is reasonable to assume that for outer planets, too, the sense of rotation on the epicycle was the wrong way.

We find support for this assumption in the elder Pliny's account of astronomical matters in the second book of his *Natural History*. Pliny's text shows many similarities to P. Mich. 149; but in this instance it would by itself offer but weak evidence, for Pliny's understanding of astronomy was faint, and his presentation consequently obscure.

Thus, there is incontrovertible evidence for models for Venus, and likely for Mars as well, with the planet moving the wrong way around on the epicycle, and we shall now proceed to show that such models must have been entirely qualitative.

Wrong-Way Epicycle

Let us now consider an epicyclic model with the wrong sense of rotation on the epicycle. We saw that in such a model the middle

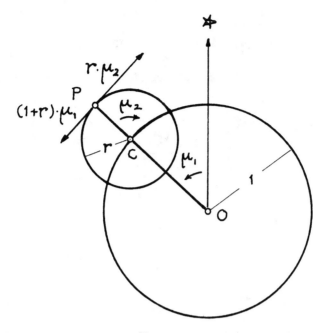

FIGURE 6.

of the retrograde arc must be reached by the planet P when it is at the apogee of the epicycle (see Figure 6).

Now let the center C of the epicycle move on the deferent so that the radius from the observer O at the deferent's center to C rotates counterclockwise about O with constant angular velocity μ_2 *relative to* OC *or its extension*. We follow Ptolemy and normalize the model and call the deferent's radius 1 (Ptolemy uses 60); measured with this unit of the epicycle's radius is r (<1). As we said before, the values of μ_1 and μ_2 are determined within rather narrow limits by simple period relations, and let them be measured with the same unit, say, radians per time unit, or rotations per time unit, or what you will—the choice of the common unit is irrelevant.

Now let the planet P be at the apogee of its epicycle (see Figure 6). The velocity of P now has two components: one in the forward direction due to the motion of OCP in the amount of

$$OP \cdot \mu_1 = (1+r) \cdot \mu_1,$$

the other in the backward or retrograde direction due to the motion of P on the epicycle in the amount of

$$CP \cdot \mu_2 = r \cdot \mu_2,$$

and the two components are along the tangent to the epicycle at P.

Now, if the planet is to be retrograde, we must clearly have that

$$r\mu_2 > (1+r) \cdot \mu_1 \tag{1}$$

or

$$\mu_2 > \left(\frac{1}{r}+1\right) \cdot \mu_1. \tag{2}$$

Since $r < 1$ and positive, $1/r > 1$, so the inequality (2) implies that

$$\mu_2 > \left(\frac{1}{r}+1\right) \cdot \mu_1 > 2\mu_1. \tag{3}$$

So if

$$\mu_2 < 2\mu_1, \tag{4}$$

an epicyclic model with the planet traveling the wrong way on the epicycle cannot possibly yield retrogradation.

For Venus and Mars we even find that $\mu_1 > \mu_2$, so the inequality (4) is amply satisfied, and thus these two planets will not become retrograde no matter what value of $r < 1$ one adopts.

Epicycles and Movable Eccenters

Ptolemy wrote his great handbook on theoretical astronomy in Alexandria circa A.D. 150. He named it *The Mathematical Systematical Treatise*, but we usually call it the *Almagest*. It is a work in 13 books (or long chapters) of which the last 5 (Books IX–XIII) are devoted to planetary theory.

Ptolemy's final planetary schemes are refined versions of epicyclic models (with the right sense of rotation on the epicycle, for which he argues in *Alm.* IX) and I shall describe them later. Here I am merely about to argue that a simple epicyclic model is a very reasonable means of accounting, in first approximation, for a planet's behavior when viewed from the earth, and I shall identify the elements of such models with their modern counterparts.

Figure 7(a) represents the standard epicyclic model of a planet with the plane of the paper being that of the ecliptic viewed from the north. The observer on the earth is at O, the direction OA points to some fixed star, the epicyclic center C revolves on a circle of radius 1 (= 60 parts) so that the line OC rotates uniformly at the rate of μ_1 degrees per day relative to the sidereally fixed direction from O to A. Finally, the planet P moves on the epicycle of center C and radius $CP = r$, so that CP rotates uniformly at the rate of μ_2 degrees per day relative to the extension of OC beyond C. We should recall here that the planet is in the middle of its retrograde arc when it is at the perigee (the point nearest the earth) of its epicycle.

Ptolemy mentions in several places a model employing a movable eccenter as a completely equivalent alternative to an epicyclic model. His most detailed account of these matters is found in the *Almagest*, Chapter 1 of Book XII, which he devotes to the theory of retrograde motion. It is here that he gives Apollonius of Perga credit for a beautiful, but far from trivial, theorem on the determination of stationary points.

One can crudely and briefly say that a model with a movable eccenter is an epicyclic model with a small deferent and a large epicycle. I have drawn such a model in Figure 7(b)—in fact, it's equivalent to the epicyclic model in Figure 7(a). The planet P in Figure 7(b) moves on the large circle of center C' different from the observer O—hence *eccenter*—while C' in turn moves on the small circle of center O—hence *movable* eccenter.

In order to transform the epicyclic model in Figure 7(a) into its equivalent in Figure 7(b), we must complete the parallelogram spanned by OC and CP by adding C'. We observe that

$$OC' = CP = r,$$

so C' is always to be found on a circle of center O and radius r. Likewise, P will always be on a circle of center C' and radius

$$OC = 1.$$

By comparing angles in Figures 7(a) and (b), we learn that C' moves on the small circle—the *concenter*—so that the radius OC' rotates counterclockwise with the angular velocity $\mu_1 + \mu_2$ relative to the direction from O to the same fixed star we used in the epicyclic model. Furthermore, P moves on the eccenter (the large circle) so that the radius $C'P$ rotates *clockwise* with the angular velocity μ_2 relative to the extension of OC' beyond C'.

2. Greek Geometrical Planetary Models

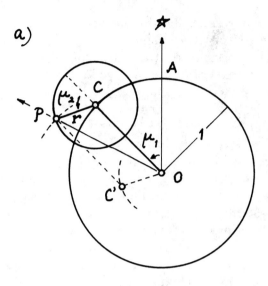

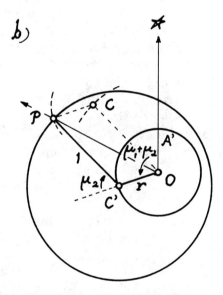

Figure 7.

If we start the motions in the two models properly, we shall afterward always obtain the same direction and the same distance from the observer O on the earth to the planet P no matter which version we use.

I shall now show by three examples how period relations of the sort we have considered in varying contexts immediately imply values of the angular velocities μ_1 and μ_2. My aim is not to reproduce Ptolemy's derivation of his refined parameters, but merely to illustrate their interdependence by considering very simple, but reasonable, relations and, most importantly, to once again emphasize the special role the sun plays in planetary models.

Let us first consider the relations for Venus:

5 synodic cycles $\approx$ 8 revolutions in the ecliptic $\approx$ 8 years.

We seek to transform these data into other units, degrees and days, and observe that one synodic cycle corresponds to one revolution of the planet on its epicycle relative to the extension of OC—the planet becomes retrograde in the neighborhood of the epicycle's perigee—and that one revolution of the synodic pattern in the ecliptic means one revolution of the epicycle's center C in the deferent relative to the direction from the observer at O to a fixed star (we recall that the plane of the paper for our models is that of the ecliptic). Additionally, one revolution in the respective circle corresponds to 360°, and for our present purpose we set

$$1 \text{ year} = 365\tfrac{1}{4} \text{ days}$$

(Ptolemy uses $365\tfrac{1}{4} - 1/300$ days). For Venus we now have

$$\mu_1 = \frac{8 \cdot 360°}{8 \cdot 365\tfrac{1}{4}\,^d} = 0;59,8,\ldots\,^{°/d}$$

[this is also the (mean) motion of the sun around the observer] and

$$\mu_2 = \frac{5 \cdot 360°}{8 \cdot 365\tfrac{1}{4}\,^d} = 0;36,57,\ldots\,^{°/d}$$

Analogously, for Mars we have

15 synodic cycles $\approx$ 17 revolutions in the ecliptic $\approx$ 32 years,

whence

80   2. Greek Geometrical Planetary Models

$$\mu_1 = \frac{17 \cdot 360°}{32 \cdot 365\frac{1}{4}\,d} = 0;31,25\ldots\,°/d$$

and

$$\mu_2 = \frac{15 \cdot 360°}{32 \cdot 365\frac{1}{4}\,d} = 0;27,43,\ldots\,°/d.$$

Third, the simplest reasonable relations for Jupiter are

11 synodic cycles $\approx$ 1 revolution in the ecliptic $\approx$ 12 years,

and so

$$\mu_1 = \frac{360°}{12 \cdot 365\frac{1}{4}\,d} = 0;4,55,\ldots\,°/d$$

and

$$\mu_2 = \frac{11 \cdot 360°}{12 \cdot 365\frac{1}{4}\,d} = 0;54,12,\ldots\,°/d.$$

For the sake of comparison, I listed in Table 1 Ptolemy's values of μ_1 and μ_2 abbreviated to two sexagesimal digits.* I appended the sums of μ_1 and μ_2 for the outer planets in the last column; all three are the (mean) daily progress of the sun.

Thus, for an inner planet, the center of its epicycle [C in Figure 7(a)] travels so that the line from the observer to it, OC, rotates with the same angular velocity as the sun or, in other words, the angle between OC and the direction from O to the sun is constant. This constant is 0, for when the planet is in the middle of its retrograde arc, it is

(i) at the perigee of its epicycle and
(ii) in inferior conjunction with the sun.

In this configuration the observer, the planet, the center of its

---

*If I had given Ptolemy's values to three or more sexagesimal places, I would have had to correct for precession, for his mean daily motion in longitude of the epicycle's center (what we shall call m_1 later) is counted from the the vernal equinox—his longitudes are tropical, as we say. However, our μ_1 describes mean motion relative to the deferent's apogee, which is sidereally fixed and so subject to precession. We should then subtract a little less than 6 in the third sexagesimal place from m_1 to get μ_1, for that is the amount of precession in degrees per day.

TABLE 1.

	μ_1	μ_2	$\mu_1 + \mu_2$
Saturn ♄	0; 2, 0,...°/d	0; 57, 7,...	0; 59, 8,...
Jupiter ♃	0; 4, 59,...	0; 54, 9,...	0; 59, 8,...
Mars ♂	0; 31, 26,...	0; 27, 41,...	0; 59, 8,...
Venus ♀	0; 59, 8,...	0; 36, 59,...	
Mercury ☿	0; 59, 8,...	3; 6, 24,...	
Sun ☉	0; 59, 8,...		

epicycle, and the sun must therefore be in a straight line. In particular, the directions from O to C and from O to the sun are equal in this situation, but then they must always remain so, for the constant must be 0, as stated. Thus, for an inner planet the line from O to C always points to the sun.

We can argue in an analogous manner that for an outer planet the epicyclic radius—CP in Figure 7(a)—is always parallel to the direction from the observer O to the sun. First we remark that the two directions differ by a constant; this is most easily seen from the movable eccenter in Figure 7(b), where OC', always parallel to CP, rotates with the angular velocity $\mu_1 + \mu_2$, which for an outer planet equals the mean daily angular progress of the sun (see Table 1). Next we observe that this constant must be 0, for when the outer planet is at the middle of its retrogradation it is also

(i) at the perigee of its epicycle and
(ii) in opposition to the sun.

Thus, the role of the sun in an epicyclic planetary model is that it guides the center of an inner planet's epicycle along the deferent, and an outer planet around its epicycle.

Comparison between Ancient and Modern Planetary Models

For a proper appreciation of the Greek planetary models it is important to recognize that they are more than a mere collection of ad-hoc devices that can reproduce the planets' apparent behavior. In fact, when appropriately scaled, they turn out to be correct representations of the planets' motions relative to the earth, in *distance* as well as in *direction*.

We shall continue for a while longer to ignore the deviations from uniform mean behavior, as we have done so far in this chapter. The modern mean model of the solar system then has the planets move uniformly, with respect to the direction from the sun to a fixed star, in circular orbits that have the sun as their common center and that all lie in one plane, that of the ecliptic. Mercury has the smallest orbit and then, in order of increasing distance from the sun, are Venus, the Earth—now a planet, though very special to us—Mars, Jupiter, and Saturn, to consider only the planets known before the 18th century (W. Herschel discovered Uranus in 1781).

Matters would then be very simple indeed if this very simplified solar system were viewed from the sun. Fortunately, we do not observe from the sun, but from the earth—a much more comfortable place to live—so our task will be to transform this heliocentric system so it displays how the planets appear to behave to an observer on the earth.

Let us first consider the case of the sun. In Figure 8(a) I have shown eight successive positions of the earth, $E_1, E_2, \ldots, E_8$, in its simplified orbit, a circle with the sun S as its center. Let us assume that the ray from S to E_1 points to a certain fixed star in the ecliptic; the eight positions are drawn so that the angle subtended at S between one of them and the next is always the same, but then the time it takes the earth to pass from one position to the next is always the same, too, for it moves uniformly in its path relative to the direction from the sun to a fixed star. The arrows from the various positions of the earth all point toward the sun and are equal in length to the radius of the earth's orbit—we call this mean distance of the earth from the sun the *astronomical unit*.

We now translate this information into a coordinate system where the earth is the origin—a geocentric system—to describe how matters appear to us. In Figure 8(b) the earth is at E, and we know that the sun is at the tip of the arrows beginning at $E_1, E_2, \ldots, E_8$ in Figure 8(a). We draw them from E and obtain the corresponding positions $S_1, S_2, \ldots, S_8$ of the sun. These eight positions lie on a circle of center E and with a radius equal to one astronomical unit. They are equally spaced in time as in direction, so the sun moves uniformly with respect to the direc-

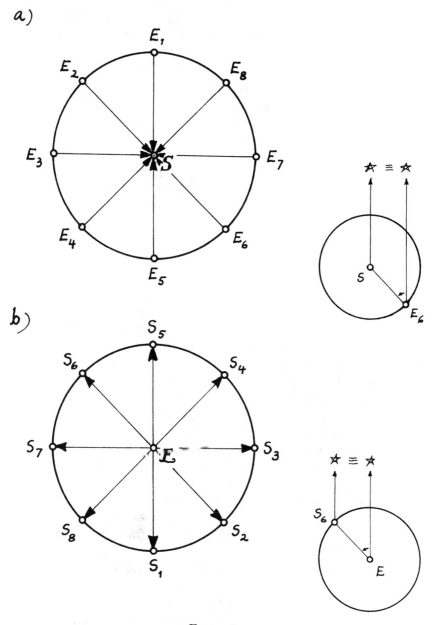

FIGURE 8.

tion from E to say, S_5 or any other direction that remains fixed with respect to ES_5. Now SE_1 in Figure 8(a) points to a certain fixed star, as we said, so ES_5 in Figure 8(b) points to the *same* star. However, this direction from the earth to the star does not change with the place of the earth in its orbit because the astronomical unit is negligible in comparison with the distance to even the nearest star. (Very strictly speaking, there is a change—we call the effect of the earth's movement on the direction from it to a star the "annual parallax of the star"—but we are dealing with quantities of less than one second of arc and so at the very limit of what is observable with even the most refined modern techniques.) Thus, the sun will move uniformly in its orbit in Figure 8(b) with respect to the direction from the earth to a fixed star—I tried to illustrate these things in the small diagrams in Figure 8.

We now turn to the problem of describing how the planets behave relative to the earth, and we shall consider the two slightly different cases presented by inner and outer planets.

Figure 9 deals with an inner planet. In Figure 9(a) we have the planet P moving in its orbit around the sun S in the center, and the earth E moves likewise, but in a larger orbit. We wish to transform this situation into a model that displays directly the two things we are interested in: the direction and length of EP. We achieve this in Figure 9(b). The sun S moves, as we have seen, in an earth-centered circle, uniformly with respect to the direction from the earth E to a fixed star, and the planet P in turn moves in its orbit, uniformly with respect to the same direction and so also with respect to the extension of ES beyond S. We recognize immediately that here we have a simple epicyclic model; the radius of the deferent is one astronomical unit, and the epicycle's radius is the planets' (mean) distance from the sun.

Figure 10 deals analogously with the case of an outer planet—Figures 10(a) and (b) present the heliocentric and geocentric pictures, respectively—but here the geocentric version turns out to be a model employing a movable eccenter. The radius of the small circle—the concenter—is the astronomical unit and that of the eccenter equals the planet's (mean) distance from the sun. If we prefer the equivalent epicyclic model (look back to Figure 7), as Ptolemy does, we have then that the radius of the deferent is the planet's distance from the sun, while the epicyclic radius is the astronomical unit.

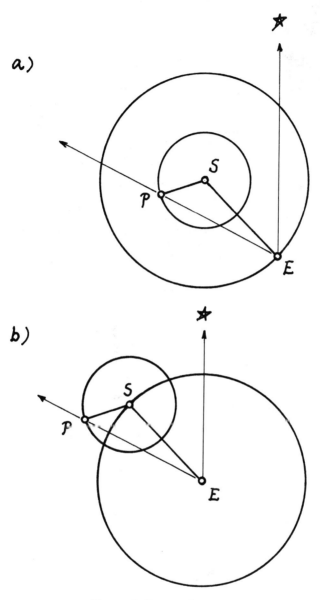

FIGURE 9. Inner planet.

86   2. Greek Geometrical Planetary Models

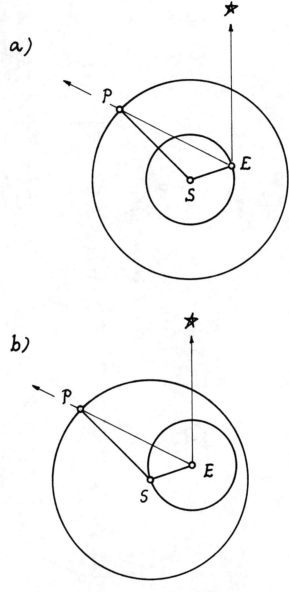

FIGURE 10. Outer planet.

In the *Almagest*, Ptolemy does not commit himself to an absolute size of his models. He is, after all, only interested in the *direction* from the earth to a planet, for that is all he can observe, and if you enlarge or shrink a model, proportionally in all parts, you will not change this direction, provided you keep the earth fixed. Ptolemy does assign absolute sizes to his models, but in another work, the *Planetary Hypotheses*, where he is concerned with cosmology, as we shall see in the next chapter.

Here I shall scale Ptolemy's models according to the correct geocentric planetary models, which we have just considered, in order to show that Ptolemy's values of the epicyclic radii are very good indeed. I wish to emphasize that in so doing, we do not violate the *Almagest* in any way, but merely add one more hypothesis.

This hypothesis is simply that the radius of the deferent for an inner planet is the astronomical unit, while for an outer planet the epicycle's radius equals the astronomical unit (see Figures 9 and 10 and recall that the concenter's radius is equal to that of the epicycle).

As said, Ptolemy always calls the radius of the deferent 1, or rather 60 ptolemaic units (p.u.) or parts. It is perfectly consistent with the *Almagest* to have the ptolemaic unit change from planet to planet, as we are about to. The epicyclic radius is r p.u. We then get for an inner planet

$$60 \text{ p.u.} = 1 \text{ a.u.} \quad \text{(astronomical unit)},$$

so

$$\text{radius of epicycle} = r \text{ p.u.} = \frac{r}{60} \text{ a.u.}$$

For an outer planet we have

$$\text{radius of epicycle} = r \text{ p.u.} = 1 \text{ a.u.},$$

so

$$\text{radius of deferent} = 60 \text{ p.u.} = \frac{60}{r} \text{ a.u.}$$

Thus, if Ptolemy's values for r are good, then $r/60$ or $60/r$ should be close to the planet's mean distance from the sun, depending on whether it is an inner or outer planet.

In the first column of Table 2 I listed Ptolemy's values for r, written sexagesimally, for the five planets. The other numbers

2. Greek Geometrical Planetary Models

TABLE 2.

	r	$\dfrac{r}{60}$	$\dfrac{60}{1r}$	a
♄	6;30	—	9.230...	9.538...
♃	11;30	—	5.217...	5.202...
♂	39;30	—	1.518...	1.523...
♀	43;10	0.719...	—	0.723...
☿	22;30	0.375...	—	0.387...

r: epicyclic radius in ptolemaic units
a: modern mean distance sun–planet in astronomical units

are written decimally, and the agreement between the values derived from Ptolemy's radii and the modern mean distances is considerable. We saw, furthermore, that Ptolemy's angular velocities (mean motions) are excellent, so the simple models as a whole are indeed very good representations of the planets' mean behavior with respect to the earth in all matters but one: namely planetary latitude.

How Ptolemy arrived at his values for radii and eccentricities —we shall introduce them later—is too tangled a tale for inclusion here. I shall only mention that the *Almagest* is in this respect no more of an autobiographical account than are most modern scientific treatises. He chooses to present a somewhat neatened version of what surely must have been a very messy process of trial and adjustment, as he himself says (*Almagest* IX, 2).

Ptolemy's Refined Planetary Models

The planetary models we have considered so far are mean models in the manner of Ptolemy. They represent the average behavior of the planets very well, as we learned, and since the underlying periods are very good, the errors inherent in all approximations do not accumulate arbitrarily, even in the course of long time intervals. In the small, however, they do not serve as well: They imply, for example, that all retrograde arcs are of equal length, and that is not so.

The Babylonians were aware of this fact and constructed their arithmetical models so they could account for it. The

Babylonians considered, at least primarily, a planet only when it was at one of its characteristic phases (for an outer planet they are first and last visibility, first and second stationary points, and opposition) and addressed the question: If a certain phase of a planet takes place at a given time and longitude, when and where will this planet's next phase of the same kind occur? Since a planetary phase happens when the planet has a certain relationship to the sun (this is most easily seen in the case of opposition), the Babylonians had only a single inequality or anomaly (the technical terms for the deviation from mean behavior) to contend with, one depending on longitude, as we saw.

Ptolemy, however, wanted his planetary models to enable him to answer the following question: Given the time, where is the planet? (We have demanded the same of our planetary theories ever since.) He had, therefore, to consider that the planet could have *any* relationship to the sun, so he had one more anomaly to account for than the Babylonians—one depending on the planet's position with respect to the sun—and that is precisely the task his epicycle enabled him to perform.

Ptolemy introduces the anomaly depending on longitude by changing the workings of the deferent in two ways. We recall that the center of the deferent played a triple role in a simple planetary model:

(i) it was the place of the observer;
(ii) it was the point from which the center of the epicycle always had the same distance; and
(iii) it was the point around which the motion of the epicycle's center was unifrom.

Ptolemy now refines such a mean model by assigning each of the three roles to a different point, thus creating what we call his *equant model* for a planet (see Figure 11).

In Figure 11 we have, as before, a deferent of center C and radius 1 ($= 60$ parts, or ptolemaic units) and an epicycle of radius r ptolemaic units with its center C' on the deferent and on which we find the planet P. However, we now observe from the point O; the motion of C' is uniform around the point Q with respect to the direction to a fixed star. The points O, C, and Q are on a straight line, the *apsidal* line, pointing to some fixed star; C is in the middle and

$$OC = CQ = e \text{ ptolemaic units} \quad (e < 60).$$

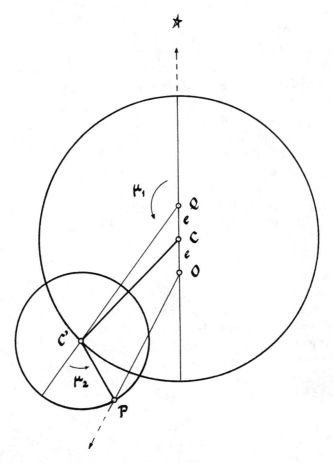

FIGURE 11.

Furthermore, the motion of P on the epicycle is uniform, not with respect to the extension of OC' beyond C' as before, but with respect to the extension of QC' beyond C'.

This is Ptolemy's simple and efficient equant model for a planet's motion in longitude—his devices for finding a planet's latitude are very complicated and will not concern us here. In the last chapter of this book, "Kepler Motion Viewed from Either Focus," I try to justify why this model is so efficient; here I merely describe how simply one may derive a planet's position from it. In this I shall not follow what Ptolemy expects a reader and user of the *Almagest* to do, although he supplies all the necessary tools, including trigonometric tables and examples of their use. Rather, he saves the reader a lot of dreary trigonometric calculations by providing auxiliary tables so that the

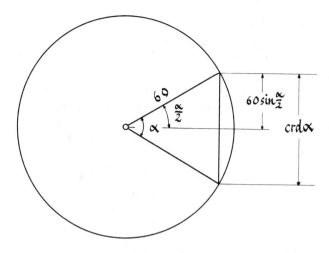

FIGURE 12.

most one has to do is interpolate in them to find corrections that are to be added to, or subtracted from, mean positions that, in turn, can be taken directly from his mean motion tables.

What I outline ahead is akin to what Ptolemy had to do when he set up his tables, and it involves solving a sequence of triangles by means of plane trigonometry.

In an earlier book, *Episodes from the Early History of Mathematics*, I presented Ptolemy's trigonometry in some detail. Here I merely recall that plane trigonometry is the branch of elementary mathematics that enables you to calculate, as accurately as you please, the measures of all the pieces of a triangle—sides as well as angles—when you are given the measures of three independent pieces: all three sides; two sides and an angle; or one side and two angles (but not three angles, for if you know two, you also know the third, since their sum is 180°). To that end we now employ a family of closely related trigonometric functions: sine, cosine, tangent, and cotangent. For a given angle, they can all be calculated with any desired precision.

Ptolemy does not use any of these but instead uses the closely related chord function: The chord of an angle α—we write crd α—is the length of the chord in a circle of radius 60 parts subtended by a central angle equal to α. We have then (see Figure 12)

$$\text{crd } \alpha = 2 \cdot 60 \cdot \sin \frac{\alpha}{2} = 120 \cdot \sin \frac{\alpha}{2}.$$

In my earlier treatment of Ptolemy's trigonometry I showed how he had calculated a table of chords of angles from 0° to 180° in steps of half a degree and how he used this table to solve all cases of triangles with three given independent pieces. We shall now see how this suffices to make his equant model for a planet yield an answer to the following question: Given the time, what is the planet's longitude?

We take for granted that the model is completely determined; that means, first, that all its dimensions are known. We have [see Figure 13(a)] the observer at O, the center of the deferent at C, the equant point Q on the extension of OC so that

$$OC = CQ = e;$$

the deferent's apogee is at A, its perigee at Π. The center of the epicycle is C', its radius is r, and the planet itself is at P. The radius of the deferent is set equal to 60 parts (60^p), and e and r are measured in such parts.

The angles that grow uniformly with time are marked α and β in Figure 13(a); in order to find them, we must know their rates of growth as well as their initial values. Ptolemy did not, in fact, find α directly, but as the difference between $\bar{\lambda}\,(C')$, the uniformly growing mean longitude of the epicyclic center C', and $\lambda\,(A)$, the longitude of the apogee A, which Ptolemy, correctly assumed to be sidereally fixed, and so subject to precession. I show, in Figure 13(a), the direction $\bar{\lambda}\,(C')$ as coming from Q, the center of uniform motion.

We must then know m_1 and m_2, the rates of growth of $\bar{\lambda}\,(C')$ and β, respectively, as well as the values $\bar{\lambda}_0\,(C')$ and β_0 of these two quantities at epoch (year 1 of Nabonassar, day 1 of month Thoth, noon), when the apogee's longitude was $\lambda_0\,(A)$. The apogee, we recall, is subject to precession, and Ptolemy's value of its rate is 1° per century. We observe that α also grows uniformly and at the rate of m_1 diminished by the rate of precession (we called the resultant angular velocity μ_1 above), but Ptolemy chooses not to make use of this.

We should futher note that for Venus

$$m_1 = 0;59,8,17,13,12,31^{\circ/d},$$

Ptolemy's value for the daily mean solar progress in (tropical) longitude, and for the outer planets

$$m_1 + m_2 = 0;59,8,17,13,12,31^{\circ/d},$$

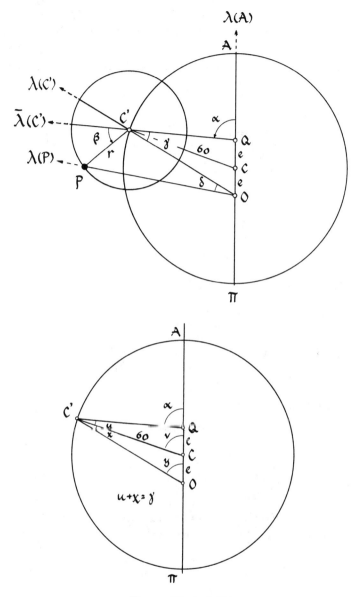

FIGURE 13. A & B

as a consequence of the special role the Sun plays in the models: For Venus, QC' points toward the mean sun, whereas for an outer planet the direction $C'P$ is the direction to the mean sun.

We are now prepared to find $\lambda(P)$, the planet's longitude at a given moment T. Our first step is to calculate t, the time interval from epoch (when $t = 0$) to T—for Ptolemy epoch is the beginning of Nabonassar's reign in -747 (746 B.C.). In the following *Excursus on Calendars and Chronology* I describe how the character of the calendars complicates the problem of finding t, how Ptolemy's *Kinglist* enables us to solve it, and the structure of the *Kinglist* itself.

Next we obtain

$$\lambda(A) = \lambda_0(A) \text{ increased by } 1° \text{ per century since epoch,}$$

and

$$\bar{\lambda}(C') = \bar{\lambda}_0(C') + m_1 \cdot t,$$

so we get

$$\alpha = \bar{\lambda}(C') - \lambda(A),$$

and furthermore,

$$\beta = \beta_0 + m_2 \cdot t.$$

These things having been given and found, we can now dispatch our problem by solving three triangles in quick succession.

First consider Figure 13(b). In $\triangle QCC'$ we have

$$\angle Q = 180° - \alpha,$$
$$CQ = e^p,$$
$$CC' = 60^p,$$

so we can calculate the angles u and v (and the side QC').

Next, in $\triangle COC'$ we now know that

$$\angle OCC' = 180° - v,$$
$$OC = e^p,$$
$$CC' = 60^p,$$

so we now can find the angles x and y and the side OC'. We set

$$\gamma = u + x.$$

Third, we turn to Figure 13(a) and consider $\triangle OC'P$, in which we now have

$$C'P = r^p,$$

OC'—just found,

$$\angle OC'P = 180° - (\beta + \gamma),$$

and so we can derive δ (and OP).

Finally, we now have for the planet's longitude

$$\lambda(P) = \lambda(A) + y + \delta$$
$$= \lambda(C') + \delta,$$

and we are done.

Excursus on Calendars and Chronology

In everyday usage, "time" can mean either of two things: a certain instant—"a point in time"—or "duration" or "time interval." A common, and important, problem in astronomy is to find the time interval between two points in time. This ought to be easy enough, but often it is not—just try to calculate how many days you have lived so far. The problem is, of course, caused by the structure of our calendar with its leap years and months of different lengths—28, 29, 30, or 31 days.

The Mesopotamian calendar was worse in this respect; as we saw, a date was given in the form

year n of King N, month x, day y,

and a moment was indicated by a number of time degrees before or after sunset or sunrise. The calendar was strictly lunar—a new month began at sunset on the evening when the new moon could first be seen, so a month was either hollow or full, that is, it was either 29 or 30 days long. Furthermore, a calendar year consisted of either 12 or 13 such months, and these years of different lengths were not distributed according to a fixed pattern (the "Metonic" cycle of 19 years) until the last 400 to 500 years B.C.

In order to find the precise number of days between two Mesopotamian dates, one would then have to know the character of all the intervening years and months. It is clear that such information was to be had, for in the *Almagest* we find Babylonian observations presented in the splendidly uniform Egyptian calendar, about which more below.

Ptolemy says that Hipparchus "arranged" the Babylonian observations, and "arranging" may well mean that he saw to it that the Babylonian dates were translated into their Egyptian equivalents. The Egyptian calendar was very simple, and therein lies its virtue. The Egyptian year was invariably 365 days long, and it was divided into 12 months of 30 days each, and 5 extra ("epagomenal") days—no exceptions, no intercalations. One may object that this year is too short, and that the calendar is therefore not "right," but there is, of course, no such thing as one, or even *a*, "right" calendar. The most important things about a calendar are that everyone within the culture agrees on it and, next, that it more or less does what you want it to do. We think it important to keep the year in step with the seasons, within a day or so, and so we have to contend with years with different numbers of days, for the tropical year is, alas, not a whole number of days long.

The Egyptian choice of a constant year of 365 days is clearly a gesture in the direction of the tropical year, but it is about a quarter-day too short. In the course of a life-span of 80 years—rarely attained in antiquity—the date of, say, the vernal equinox would then creep some 20 days forward toward higher dates through the month or months, but this was a small price to pay for the rigid regularity of the calendar, at least from an astronomer's point of view.

The year in which an event took place was, as said, given as a regnal year of the reigning king of Babylon, and so it was in Hellenistic Egypt. To make use of this information, you must obviously know when the king reigned, and you can learn that from Ptolemy's *Kinglist*, which he appended to the *Almagest*. In it he presents the sequence of successive kings, first of Babylon and then of Egypt, and the lengths of their reigns in Egyptian years, as well as a running total, beginning with Nabonassar, king of Babylon ($-746 = 747$ B.C.). (See Figure 14.)

After Ptolemy; Copernicus, Brahe

Ptolemy's *Almagest*, composed in the second century of our era, has reached us in the original Greek in Byzantine manuscripts, transferred to parchment from the less durable papyrus. The printed editions of the Greek text—the best and most recent was delivered by the Danish classical scholar J. L. Heiberg from 1898 to 1903—as well as the English translation by G. J. Toomer (1984) are based on such manuscripts from the ninth and tenth centuries.

The *Almagest* remained the central astronomical treatise for nearly a millennium and a half whereever serious positional astronomy was cultivated, also outside the Greek-speaking domains, first in the Islamic world and later in medieval and Renaissance Europe. That is not to say that it was slavishly accepted and passed on. Indeed, we find both improved parameters and alternative geometrical models suggested by later astronomers, but the *Almagest* remained the touchstone against which they tested such models.

Indeed, around A.D. 1000 Islamic astronomers began subjecting the *Almagest* to constructive criticism and emendation, on both astronomical and philosophical grounds—it had been translated several times into Syriac and Arabic in the late eighth and ninth centuries. It was a Latin translation of one of these Arabic versions, made in Toledo in 1175 by Gerard of Cremona, that in effect introduced the *Almagest* to the West; earlier Latin translations from the Greek had seen scant use. What Ptolemy called *The Mathematical Systematic Treatise (Syntaxis)* now got the Latin name *Almagesti* or *Almagestum*, a transliteration of an Arabic term that, in turn, probably was a transliteration of the Greek *megiste (syntaxis)*, meaning *"the greatest (treatise),"* though this appropriate title is found in none of the surviving Greek manuscripts.

I give but a few examples of such emendations and criticisms; simplest among them is the introduction of improved values of parameters, particularly those that measure extremely slow rates of change, for to determine such slow changes well, one must observe and keep records over long time intervals. Indeed, Ptolemy was well aware of this and invited his successors to improve what he himself had had to be satisfied with. Notable among such slow changes is the "precession of the equinoxes," the extremely slow increase in (tropical) longitude of the fixed stars. Hipparchus, who discovered the phenomenon, had estimated it to be at least 1° per century, and Ptolemy adopted that value, a nice round number, easy to remember, but too low—the correct rate is 1° in some 71.6 years. Islamic astronomers proposed values closer to the modern one.

A related problem concerns the apogee of the solar model. Hipparchus described the behavior of the sun by a simple eccen-

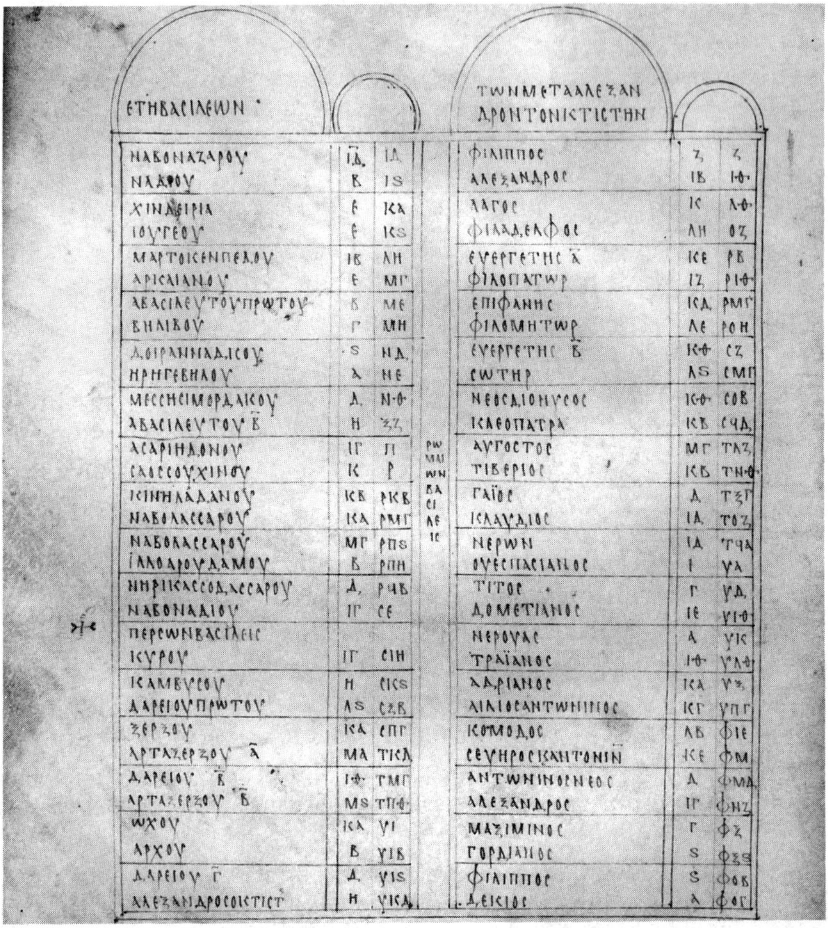

FIGURE 14. Ptolemy's *Kinglist*. Figure 14 shows Ptolemy's *Kinglist* as it appears on folio 16, verso, of the Byzantine manuscript Vaticanus graecus 1291, written in the first half of the ninth century. The manuscript contains otherwise a version of Ptolemy's *Handy Tables*, his elaboration and extension of the *Almagest*'s tables, occasionally with altered parameters.

The first set of three columns of the *Kinglist* is headed "Years of the Kings." The first 20 lines concern 18 kings of Babylon in order of their consecutive reigns, beginning with Nabonassar and ending with Nabonaides, and two interregna. The next 10 lines have the subheading "Persian Kings" and list the rulers of the Persian dynasty, beginning with Cyrus and ending with Dareius III; in the last line we find Alexander "the Founder."

The second set of columns is headed "[Years] of those after

tric model; it has the sun moving in a circle, uniformly around the center, which is outside the earth. He assumes that the apogee—the point where the sun is farthest from the earth—is tropically fixed, that is, its longitude measured from the vernal equinox remains constant, and Ptolemy adopted Hipparchus's

---

Alexander the Founder" and begins with Alexander's two successors as king of Babylon, Philip (Arrhidaeus) and Alexander (the younger). Then the list changes to the Macedonian rulers of Egypt, all but the last called Ptolemy, and here they are given only their cognomina, Lagos, Philadelphos, Euergetes I, Philopater, etc. Ptolemy I, once Alexander's Companion in Arms, and founder of the dynasty, is here identified as Lagos, his father's name, and not as Soter, his more usual cognomen. The last of the line is Cleopatra and then, as noted in the margin, the list gives the names of the Roman "kings," Augustus, Tiberius, Gaius, and so forth.

The second column of each set records the length of each reign, in Egyptian years of 365 days, and the third keeps a running total, beginning with Nabonassar's reign. In this version of the *Kinglist*, the total ends at 424, including Alexander's 8 years, and begins afresh with Philip's reign—not only was an Era Nabonassar in use, but also an Era Philip. [Year 1 of Era Nabonassar, month Thoth of the Egyptian calendar, day 1 corresponds to −746 (= 747 B.C.), February 26.]

Ptolemy's *Kinglist* or *Canon* was long the principal key to ancient chronology, and the Babylonian part has on the whole been confirmed by the cuneiform evidence as it became available in the course of the last century and a half.

Those who would enjoy reading what was written some 1200 years ago might need to know the numerical values assigned to the Greek letters. They are given in Table 3, where the classical Greek alphabet has been augmented by three obsolete letters: digamma (the old waw), also called stigma, for 6; qoppa for 90; and sampi for 900.

The first two lines say

NABONAZAPOY	IΔ	IΔ
NAΔPOY	B	IS

which means

of Nabonassar	14	14
of Nadros	2	16.

For a recent reconstruction of the *Kinglist*, see G. J. Toomer's *Ptolemy's Almagest*, p. 11.

TABLE 3.

A	α	1	I	ι	10	P	ρ	100
B	β	2	K	κ	20	Σ	σ	200
Γ	γ	3	Λ	λ	30	T	τ	300
Δ	δ	4	M	μ	40	Y	υ	400
E	ε	5	N	ν	50	Φ	φ	500
F	ϛ	6	Ξ	ξ	60	X	χ	600
Z	ζ	7	O	ο	70	Ψ	ψ	700
H	η	8	Π	π	80	Ω	ω	800
Θ	θ	9	Ϙ	ϙ	90	ϡ	ϡ	900

ʘ, ȣ, etc. = 0

model without change. Several early Islamic astronomers realized that the solar apogee is certainly subject to precession, just like the planetary apogees, and some, at least as early as al-Biruni (ca. A.D. 1025), even detected a further proper motion.

Two further examples will suffice, but they concern more substantial and serious criticism of Ptolemy's work. In one of them the criticism is justified, in the other not. The first concerns his lunar model, which, alas, is too complicated for inclusion here. I shall merely say that in order to account for the fact that lunar anomaly (the moon's deviation from uniform behavior) is larger at quadrature than at syzygy, Ptolemy introduces a crank mechanism in his model that pulls the epicycle nearer at quadrature to make it appear larger—indeed, nearly twice as large. Ptolemy conveniently does not mention that the moon itself consequently ought to look twice as large at quadrature than at syzygy (when it is new or full), and anyone can see it does not. This serious flaw was removed by the replacement of the crank linkage by other geometrical devices, one of which—the Ṭūsī couple—I mention later.

The other criticized deficiency was considered much more fundamental: When he introduced the equant model, Ptolemy committed an act of philosophical heresy. Aristotle had argued that the only self-sustaining natural motion was uniform circular motion, and his argument had been so effective that natural philosophers, including Copernicus, remained persuaded for nearly two millennia. Ptolemy himself pays his respects to the principle of uniform circular motion in the preface to the *Almagest*, but he blithely violates it when he constructs satisfactory models for the moon and planets. To be sure, the motion on the equant model's deferent is both circular and uniform, but it is

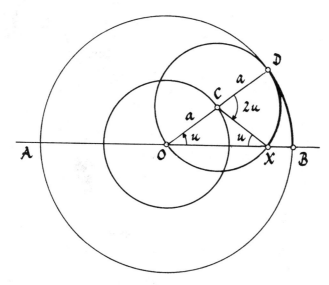

FIGURE 15.

not uniform around the circle's center, and so not of constant speed, and that would not do for a proper Aristotelian.

Medieval Islamic scholars expended much effort and ingenuity on the construction of models in which combinations of truly uniform circular motions very nearly did the same job as the the equant model. Indeed, there was never any question about the equant model's efficiency, and the test for a philosophically correct model remained that it produce nearly the same positions as an equant model.

The earlier center for astronomical activities of this kind was at Marāgha in northwest Iran with Naṣīr al-Dīn al-Ṭūsī (1201–1274) and his school. In the 14th century the brilliant Ibn al-Shāṭir of Damascus wrote a treatise proposing models that were composed entirely of uniform motions and that worked as well as Ptolemy's.

One important component of these structures is what we now call a "Ṭūsī couple" after its inventor in Marāgha. This lovely and simple geometrical device produces, surprisingly, a harmonic oscillation in a straight line as a result of only two combined uniform circular motions and can be presented in two equivalent forms.

In Figure 15 let A, O, and B be points on a fixed straight line such that

$$AO = OB = 2a.$$

Consider now the circle with center O and radius a and another circle with its center C on the circumference of the first circle and the same radius a. Let X be one intersection between the second circle and the line AB (the other is O) and draw the diameter OCD in the second circle. Now $\triangle OCX$ is isosceles (with two sides equal to a), so the angles at O and X are equal, say u, and they are so marked. Consequently, we always have

$$\angle XCD = 2u,$$

for it is exterior to the third angle in the triangle and thus equal to the sum of the first two (this is the same as saying that the sum of the angles in a triangle is 180°).

We now use this bit of geometry when we devise a kinematic model: We consider the first circle as a deferent and the second circle as an equally large epicycle. We let the deferent's radius OC rotate counterclockwise around O, uniformly with respect to the fixed direction OB, namely, we let u grow uniformly; additionally, we let the epicyclic radius CX rotate in the opposite direction—clockwise—uniformly with respect to the extension CD of OC, but with twice the angular velocity of u. If we start the motion with $u = 0$—so X is at B—X will move from B to A and back again, and the oscillation will be harmonic, for we have

$$OX = 2a \cdot \cos u$$

(with sign).

This is a Ṭūsī couple, and it produces a harmonic oscillation in a straight line in a way that could not offend even the sternest Aristotelian eye, namely, as the result of combining two uniform circular motions, in opposite directions and the second twice as fast as the first.

We can also interpret Figure 15 in a different and "mechanical" way: We draw the circle with O as a center and $OB = 2a$ as a radius, and we note that

$$\text{arc } BD = \text{arc } XD = 2a \cdot u$$

if u is measured in radians. Thus, a point on the circumference of a circle will describe a rectilinear, harmonic oscillation when the circle rolls without slipping, and with constant speed, inside another circle that is twice as large.

If we place such a Ṭūsī couple in the plane of the ecliptic with the diameter AB perpendicular to the line of sight, we make the point X oscillate harmonically and we have thus gained the following two advantages: First, the motion is generated entirely by means of uniform circular motions; second, the point X remains at practically the same distance from us.

Similarly, if we see to it that the couple's diameter AB always points to the observer, or some other fixed point, we can make the distance from the fixed point to the point X vary periodically without interfering with the direction to it.

Thus, the Ṭūsī couple enables us to vary *either* the direction *or* the distance to an object, whereas with the old device of an epicycle it is either both or nothing, and this makes it a very useful and versatile component of philosophically correct models.

An important example of what can be achieved with this device is a result by Naṣīr al-Dīn al-Ṭūsī and two of his followers. Using a combination of only uniform circular motions, they succeeded in making the center of the epicycle move uniformly around the old Ptolemaic equant point—*exactly*, not approximately—while the new orbit of the epicyclic center stays very close to the old deferent.

I shall be more specific. In Figure 16 I show a Ptolemaic deferent of center C and radius ρ. The observer is at O, and the equant point is Q. The eccentricity—much exaggerated—is e (here I use e to indicate the length OC, not a ratio, so $e = \varepsilon\rho$, where ε is what we usually mean by eccentricity) and Λ is the apogee, Π the perigee. The epicycle's center C' moves uniformly around Q (i.e., the angle α grows uniformly with time) and not around the center C; this is what was considered so philosophically offensive. Furthermore, I display, following E. S. Kennedy, three equivalent attempts at doing nearly as well as Ptolemy, but with nothing but uniform circular motions.

The first is by Naṣīr al-Dīn al-Ṭūsī himself. He extends the line QC' to D so that $QD = \rho$, the deferent's radius. He then places a Ṭūsī-couple with its center at D and principal diameter along QD and its extension, and with the radius of the small circle equal to $e/2$. If the angles are as shown in the figure, C''—the equivalent of X in the preceding discussion of the device—is his epicyclic center. Naṣīr al-Dīn uses the version of his couple with a circle rolling inside another that is twice as large. I show the arrangement in Figure 17. Since C'' lies on the line QD, it moves uniformly around Q when α grows uniformly with time, precisely like Ptolemy's C'.

We should further note that the radius of the small circle is

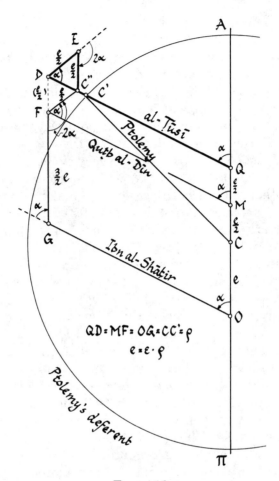

FIGURE 16.

chosen so that C'' coincides with the apogee A and the perigee Π when α equals $0°$ and $180°$. This ensures that C'' will not stray far from the old deferent, particularly when e is small.

In Figure 16 I also indicate two almost identical other ways of reaching the point C'' via uniform circular motions (for the sake of clarity, I omitted the circles themselves and drew only the relevant radii). The first variant is due to Quṭb al-Dīn al-Shīrāzī (1236–1311), who manages with only two circles. The first has its center at M, the midpoint of CQ, and radius ρ; and the second has its center at F, where MF is parallel and equal to QD, and radius $FC'' = e/2$. The motions are as indicated in Figure 16.

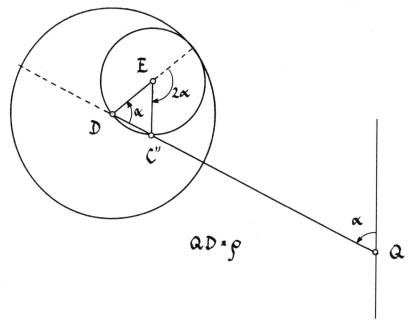

FIGURE 17.

The second version is from the work of Ibn al-Shāṭir of Damascus (1304–1375/6). His model is quite like Quṭb al-Dīn's, but instead of going from O to F via M, Ibn al-Shāṭir goes via G, where G is the fourth vertex in the parallelogram spanned by O, M, and F. Quṭb al-Dīn needed only two circular motions, the first of which had the eccentricity $3/2\ e$, while Ibn al-Shāṭir needed three, but they are without eccentricity.

Ibn al-Shāṭir's model is found in his treatise, nobly entitled *The Final Quest Concerning the Rectification of Principles*, in which he also makes ample further use of al-Ṭūsī's device. In the 1950s E. S. Kennedy rediscovered this treatise, and we were astounded to learn that component for component, link for link, Ibn al-Shāṭir's models correspond to those of Copernicus, perhaps with the order of the links changed.

Of Copernicus and Brahe

I shall now conclude this chapter on Greek geometrical models with a brief sketch of parts of the work of Copernicus and Brahe.

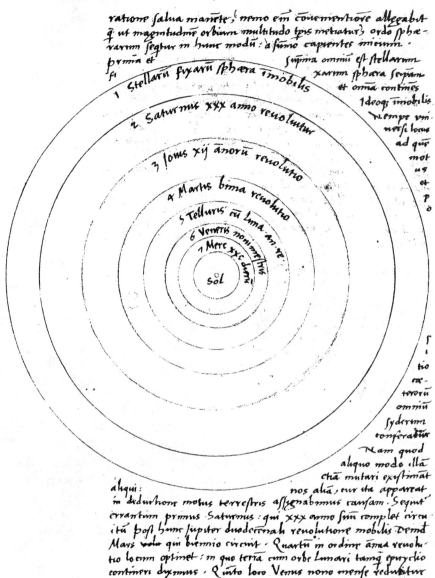

FIGURE 18. This plate shows the Copernican arrangement of the universe as Copernicus drew it himself (not to scale). Outermost we see the immobile sphere of the fixed stars, inside which are the planetary orbits (or spheres), among them the Earth's, and the Sun in the middle. The orbits' legends include approximate periods: Saturn 30 years, Jupiter 12 years, Mars 2 years, the Earth with the Moon 1 year, Venus 9 months, and Mercury 80 days.

The figure is part of a manuscript copy of *De revolutionibus* ... in Copernicus's own hand. He gave it to his student Rheticus. After

This is quite proper, for though they in a sense marked the beginning of modern astronomy—Copernicus with his cosmology, Brahe with his vast corpus of the best observations yet—their theoretical work was very much in the ancient tradition.

But first I cannot help remarking that if one is accustomed to working on ancient astronomy, a field where one must stretch to the utmost every fragment of primary evidence, be it of clay tablet or papyrus, and where treatises must be laboriously reconstructed from copies of copies, many times, and centuries, removed, one becomes entirely overawed by the wealth of source material preserved from the Renaissance. We have correspondences, journals, all sorts of records, contemporary printed editions (Brahe had his own printing press and papermill on Hveen), and even a manuscript copy of *De revolutionibus*... (see Figure 18) written very beautifully by Copernicus himself in red and black ink (it is now in Krakow and in pristine condition, for it is not the copy used by the printer).

Thus, we have quantities of information about Copernicus and Brahe, and accounts of their lives can be found in many places, and in varying degrees of detail, so I provide just the briefest biographical sketches of them.

Nicolaus Copernicus was born in 1473 in Torun (Thorn), Poland, in his father's handsome house that still stands inside the medieval city walls. He died in 1543 in Frauenburg (Fram bork) two months after the publication of his principal work, *De revolutionibus orbium coelestium* ("On the revolutions of the celestial spheres"). He attended the University of Krakow and afterward studied medicine and canon law (and astronomy on the side) in Italy. He spent most of his life as canon in the Chapter of Warmia, where his uncle and protector, Lucas Watzenrode, was bishop from 1489 to 1512.

De revolutionibus ... is in its structure, compass, and mathe-

---

changing ownership many times, it is now in the Jagiellonian Library in Krakow.

The whole manuscript has been published in facsimile twice, first by Verlag R. Oldenbourg, Berlin-Munich, in 1944 (this figure is copied from this edition) and again by the Polish Academy of Sciences in Warsaw in 1975.

matical and astronomical methods very like the *Almagest* except in two respects: It includes cosmology, and it adheres strictly to the principle of uniform circular motion. Copernicus's cosmology, based on the heliocentric hypothesis, is, of course, what he is famous for. I shall later compare it to Tycho Brahe's geocentric version in the context of the astronomy of the time.

As an illustration of the second point of difference, I can mention that Copernicus used two of the three models in Figure 16, with obvious modifications. First, in the *Commentariolus*, a short, early presentation of his principal ideas that was circulated in manuscript, we find Ibn al-Shāṭir's model from Figure 16, but with the sun at O and the planet at C''. In *De revolutionibus* ... itself, perhaps to save one circle, Copernicus replaced it with Quṭb al-Dīn's obvious alternative, as I show in Figure 19. Here S is the sun, P' the planet according to both of Copernicus's models, and P where the planet would be according to a Ptolemaic equant model.

The degree of agreement between Copernicus's models and those of Ibn al-Shāṭir is in general such that independent discovery is quite out of the question, but we do not know where, and in what form, Copernicus learned of the earlier work. It has often been claimed that Copernicus wanted to get rid of the equant point. This is wrong. In fact, he strove to preserve it, but in a philosophically correct manner, and he succeeded in doing so by following the work of the Marāgha School, as we have seen.

Tyge (in Latin, Tycho) Brahe (1546–1601) was born into a noble Danish family—a Rosenkrans and a Guldensteren (Brahe's spellings) are among the ancestors whose coats of arms he displays around his portrait—at his family's seat Knudstrup in Skåne, since 1658 a Swedish province. He spent most of his earlier adulthood traveling in Europe, getting acquainted with leading astronomers and instrument makers concerned with improving instruments and observational techniques. He made observations that gained him such renown that when he was 30 years old, his king, Frederik II, gave him free use for life of the island Hveen—also Swedish since 1658—and several benefices whose proceeds he used to build, maintain, and adorn his observatory Uranienborg and associated workshops. Until he quarelled with the government of the young Christian IV and went into exile in 1597, he and his staff made and recorded a body of observations that became the basis of modern astronomy.

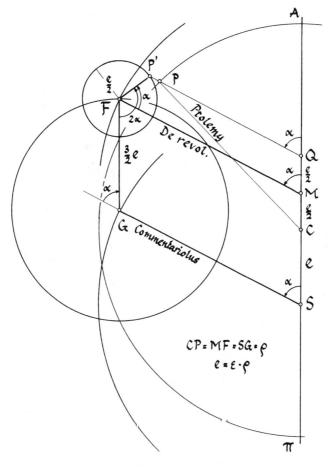

FIGURE 19.

With his best instruments he reached an accuracy of less than one minute of arc, by far the best until that time, and probably near the limit of what is achievable without the telescope. Further, and most importantly, he departed radically from the practice of his predecessors in his observational program. Hitherto one had mostly observed, say, planets when they were in particulaly interesting situations, well suited to yield parameters of preconceived models. However, Brahe and his staff ran sustained sequences of indiscriminate observations without regard to hypotheses or models. When Brahe died in exile in Prague in 1601, his observations remained at the disposal of Johannes

110   2. Greek Geometrical Planetary Models

Kepler, who in their wealth could find whatever he needed for the construction of his new theories and models.

Brahe proposed an alternative cosmological scheme to that of Copernicus—he modestly called it the Tychonic system, and so do we—and I shall present and discuss the two systems together, in qualitative terms.

The name of Copernicus immediately calls to mind his heliocentric cosmological scheme: The (mean) sun is at rest in the "center" and the planets travel in orbits around it, among them the earth, spinning on its axis. His reason for proposing a theory that seems to be at such odds with everyday experience, and common sense, lies in the peculiar role the sun played in Ptolemy's planetary models: For an inner planet, the radius of the deferent points to the mean sun, whereas for an outer planet, the epicyclic radius to the planet is parallel to the direction from the earth to the mean sun. Indeed, these seeming peculiarities are immediate consequences of the heliocentric hypothesis. I must emphasize that once one has suggested a heliocentric solar system, one can immediately find its dimensions in astronomical units from the *Almagest*'s parameters: An inner planet's distance from the sun is given by $r/60$, and an outer planet's by $60/r$, where r is the planet's epicyclic radius measured in parts (Ptolemaic units), one part being one-sixtieth of the deferent's radius, as we saw earlier (page 88).

The Tychonic system is the geocentric equivalent of Copernicus's: The earth is at rest, the sun revolves around it once a

FIGURE 20. The frontispiece to Giambattista Riccioli, *Almagestum Novum* ..., Bononia (Bologna), 1651. Riccioli (1598–1671) was a Jesuit scholar and so preferred the Tychonic system. He found it quite compatible with the recent discoveries in observational astronomy, and there are many indications of his awareness of them among the significant elements in the allegorical picture, including a telescope, Mercury and Venus showing crescents, Saturn with its rings (though they look more like ears, as they did at the first sightings), Jupiter with its moons, and the rough surface of the Moon.

Urania weighs the Tychonic against the Copernican system and finds the latter wanting. The Ptolemaic system lies on the ground under the Tychonic. A reclining Ptolemy says *"Erigor dum Corrigor,"* which may be translated as "I am exonerated even as I am corrected" —Riccioli knew, of course, that the *Almagest* is compatible with the Tychonic system.

year, and the five planets travel around the sun. The Tychonic planetary models are then very like the properly scaled Ptolemaic models, in movable eccenter version for the outer planets, that we discussed previously.

It cannot be overstressed that the Copernican and Tychonic arrangements yield identical results as far as directions and distances within the solar system are concerned, and it takes evidence of a different sort to distinguish between them. For example, the telescopic discovery of the phases of Venus is often quoted as proof of the Copernican hypothesis. It is no such thing: It merely shows that Venus goes around the Sun, *not* that the Earth does, and it is in agreement with the Tychonic system as well.

Furthermore, as soon as we try to calculate planetary positions, we are, in a sense, working in a system of the Tychonic type, for we are concerned with how things look from the earth, not from the sun. The *Nautical Almanac* is, as it were, produced on a Tychonic hypothesis.

One had, then, to seek beyond the solar system for observational confirmation of the Copernican system, and it was slow in coming. It had been clear ever since antiquity that a heliocentric system would imply an apparent yearly change in stellar positions as we on the earth travel in a vast orbit around the sun, and no such yearly parallax was observed until Bessel did so in the 1830s.

The problem of a heliocentric hypothesis combined with a failure to observe an annual parallax is nicely put by Archimedes (d. 212 B.C.) in the preface to his *Sandreckoner* (Heath's translation):

But Aristarchus of Samos (early third century B.C.) brought out a book consisting of some hypotheses, in which the premises lead to the result that the universe is many times greater than that now so called. His hypotheses are that the fixed stars and the sun remain unmoved, that the earth revolves about the sun in the circumference of a circle, the sun lying in the middle of the orbit, and that the sphere of the fixed stars, situated about the same centre as the sun, is so great that the circle in which he supposes the earth to revolve bears such a proportion to the distance to the fixed stars as the centre of the sphere bears to its surface.

Brahe had tried in vain to observe an annual parallax and argued that either there was not any, in which case Copernicus was wrong, and his own system was to be preferred, or it was less than what he could observe. He knew that his accuracy of observation was $1'$, which then had to be an upper limit on the annual parallax. This, in turn, placed a lower limit on the distance to the fixed stars. Now, he believed to have measured the angular diameters of stars to be between $1/3'$ and $2'$, and a star of diameter, say, $1'$ would be as large as the orbit of the earth, which he found absurd. (Note that a star's annual parallax can also be defined as the angle subtended by the earth's orbit at the eye of an observer on the star.) By the way, Brahe's estimate of the apparent size of a fixed star was based on an optical effect (diffraction) that makes a point source of light seem larger the brighter it is; telescopes showed that no matter how large the magnification, fixed stars still appear as points.

After the introduction of telescopes, astronomers eagerly sought the annual parallax of stars but did so in vain until F. W. Bessel in the years 1837–1840 found the annual parallax of 61 Cygni to be $0.35''$.

As a matter of fact, another, but unforeseen, consequence of the earth's motion had been observed by Bradley about a century earlier. It turned out to be what we call the "annual aberration," the apparent deflection of the line of sight to a star in the direction of the earth's velocity, which, to be sure, is small in comparison to the velocity of light, but not negligible. The effect is rather like what you observe when you drive through rain falling vertically down: The rain seems to come slanting down toward your windshield, falling from a point ahead of the one directly overhead. The aberration is about $20.5''$ for a star near the pole of the ecliptic. It should be noted that the annual parallax is smaller the farther the fixed star is from us, whereas the aberration is independent of distance.

By the time of its strict observational confirmation, the Copernican system had, of course, long been generally accepted, due particularly to the work of Kepler and Newton. Keplerian motion is the subject of the last chapter of this book, but Newton's dynamics and laws of motion are beyond our scope.

3
Ptolemy's Cosmology

Textual Problems

The Ptolemaic system has long been in common use as the name for the cosmological scheme that was eventually replaced by the Copernican. It consists of seven nested spherical shells, one for the Sun, one for the Moon, and one for each of the five planets, all surrounded by the sphere of the fixed stars. All of this revolves once a day about a central, stationary Earth, while Sun, Moon, and planets move appropriately and much more slowly, each within its own sphere. This scheme is most often represented graphically by its intersection with the plane of the ecliptic (the Sun's orbit)—it then becomes a system of nested annuli—for the planets and the Moon are always close to this plane.

In modern times the Ptolemaic system is usually encountered only when it is being compared—not to its advantage—to the Copernican scheme by authors and teachers who, more often than not, are wholly ignorant of its finer structure, dimensions, and underlying principles. For pedagogical reasons the virtues of Copernicus's system and the flaws of Ptolemy's are equally emphasized, and one is left with the impression that anyone but an utter idiot ought to have embraced Copernicanism immediately after having been exposed to it and discarded with scorn the older views.

I find it a pity that a vague and skewed image is thus created of the cosmological system that prevailed for nearly a millennium and a half in the West, and for longer in the Near East, a system of pleasing harmony and inner consistency, and with all its parts neatly measured in terrestrial measure. To give but one example of the confident and matter-of-fact manner in which medieval authors approach their universe, I can point to the place in Dante's *Convivio* where he says that "Mercury is the smallest star in the heavens, for the quantity of its diameter is not more than 232 miles according to what Alfraganius (al-Farghani) posits, who says it to be 1/28 of Earth's diameter which is 6500 miles."*

In the following I set forth the basic assumptions, explicit as well as implicit, on which the Ptolemaic system rests, and sketch how its details are derived from them, but I begin by tracing the difficult path that led to our present control of the evidence for the system's genesis.

It seems a tautology that the Ptolemaic system is due to the astronomer Ptolemy, or Claudios Ptolemaios, of Alexandria, whose principal work on astronomy, the *Mathematical Collection* or the *Almagest*, as it is more commonly called, was written about A.D. 150. Mathematicians, and even more so historians of mathematics, know all too well, however, that the person whose name is affixed to a theorem is not necessarily the one who first stated and proved it, and it seemed as if the Ptolemaic system was a case in point. Indeed, the few who had read Ptolemy were well aware that this system is nowhere described in any of his works that have survived in the original Greek, and some of these few—I for one, alas—were willing to argue, quite persuasively, we thought, that the system was not his.

Our earliest textual evidence for it was found in the *Hypotyposis* ("Outline") *of Astronomical Hypotheses* ("Models") by Proclus (A.D. 412–485). Proclus was the next-to-last director of Plato's old Academy in Athens and was passionately devoted to the old pagan Greek culture that was being threatened on all sides by the new Christian barbarism, as he thought it. You get the feeling when you read his works that he saw himself as one of the last outposts of the culture he loved so well, and he did what he could to preserve it for posterity by writing highly competent summaries of most branches of Greek learning. His works—the survivors of which, by the way, are still not completely edited—were indeed, he would have been gratified to know, much studied and admired during the early Renaissance. The text of his *Hypotyposis* was fortunately well published with a German translation by Karl Manitius in 1909, and it remains

---

*I owe this reference to Noel Swerdlow.

the best elementary introduction to classical astronomy in the Ptolemaic manner. In most of this work Proclus summarizes the contents of the *Almagest*, but he also devotes a chapter to a description of the cosmological system of nested spheres, which has no counterpart in the *Almagest*. It is curious, and seemed significant, that Proclus does not mention Ptolemy anywhere in this part, while elsewhere he is lavish with both credit and praise.

It is quite natural that a cosmological scheme is nowhere to be found in the *Almagest*, for Ptolemy's concern here is with positional astronomy and eclipse theory. In other words, the principal aim of the *Almagest* is to enable you to answer the question: Given your location on the earth, and given the time, in precisely which direction should you look in order to see a given celestial body? For only one of the bodies—the moon—is its distance from the earth of practical interest in this connection, and tables of the moon's daily parallax are duly given (they matter most for the prediction of the visibility and magnitude of solar eclipses at a given place).

The obvious place for Ptolemy to present his cosmological system, we thought, was in his *Planetary Hypotheses* (the Greek word "hypotheses" is here best translated by "models") and it was not there. Our best edition was that delivered in 1907 by the Danish classical philologist J. L. Heiberg (to whom we owe so many definitive editions of Greek mathematical and astronomical texts) in a volume presenting Ptolemy's minor astronomical works. He gave all that survives in Greek (18 printed pages), which he took to be Book I. There are, however, two manuscripts of an Arabic translation of the *Hypotheses*, one in Leiden and the other in the British Museum, and both contain Book II as well as Book I. The Arabic manuscript in the British Museum was written in 1242, while none of Heiberg's Greek manuscripts antedates the 14th century. The brilliant young German Orientalist Ludwig Nix translated the Arabic text into German, but he died before putting the finishing touches on his work. Heiberg decided to print on facing pages Nix's translation via the Arabic of his Greek text as far as it went. Heegaard and Buhl (a mathematician and an orientalist) prepared Nix's translation of Book II for publication—they had the use of the Leiden manuscript—and Heiberg printed it after the Greek and German of the first part. The work as it then stood dealt with the construction of planetary models for motion in longitude and latitude, largely based on the *Almagest*'s parameters, but nowhere did it give instructions for fitting the models for the different planets together into a cosmological whole.

This is where matters rested until 1964 when the German scholar Willy Hartner published the results of his close study of Proclus's *Hypotyposes* and commentary to Plato's *Timaeus* and of various Arabic treatises on cosmology. He came to the conclusion that the Ptolemaic system must have been set forth by Ptolemy himself in the *Planetary Hypotheses*, very likely at the end of Book II.

A few years later Noel Swerdlow was beginning to work here at Yale on a doctoral thesis on medieval cosmology, and very naturally we were discussing Hartner's article. This prompted my then-colleague, Bernard Goldstein, who directed Mr. Swerdlow's thesis, to wonder if the missing part might not be found in a Hebrew version of the *Planetary Hypotheses* even though it was absent, so it seemed, in the Arabic. He consulted Steinschneider's work on medieval translations into Hebrew and learned that he had, in fact, in his desk drawer a microfilm that among other things contained the wanted work. To his great and good surprise he found that where the Greek (and German) of Book I gave out, the Hebrew continued with Ptolemy's own account of the system of nested spheres. Goldstein obtained microfilms of the two Arabic manuscripts and found, again to his surprise, that they, too, had the same long passage in the same place—one cannot understand how it came to be omitted in Heiberg's edition. Goldstein published the complete Arabic text, with commentary and a translation of the newly found second part of Book I, in 1967. At long last it had been firmly established that the Ptolemaic system was indeed due to Ptolemy.

I have gone into these matters at such length to convey a sense of the vagaries, and the pleasures, of scholarship, and of the importance of consulting the texts themselves.

Principles, Assumptions, and Construction of the Ptolemaic System

I now finally turn to a statement and discussion of the principles and assumptions that underlie Ptolemy's cosmological scheme, and then I shall sketch how he derives it from them.

118   3. Ptolemy's Cosmology

First Ptolemy assumes that the fixed-star heaven is spherical and moves like a sphere around a spherical earth, very small in comparison and at rest in the center. He argues for all these things in the *Almagest*, Book I, and they are of course not new with him. Here I might insert a remark on the recognition of the earth's sphericity. In the mid-third century B.C. Eratosthenes had given a good estimate of the earth's circumference, and at least from that time on no person educated in the classical tradition, be it in antiquity, in the Islamic countries, or in the Latin West, had doubted that the earth is round. The popular notion that Columbus proved it is utterly false; he merely advocated a value of its size—whether he believed it or not—that was much too small and used it to argue for a shorter western route to China. As everyone knows, he never got there.

Ptolemy does not argue for the next fundamental principle, but merely takes it for granted. It amounts to this: Each of the planets, including Sun and Moon, has its own particular range of possible distances from the Earth; in other words, once a planet has been at a certain distance, no other planet can ever attain that same distance, except perhaps at the extremes of its range. It follows that each planet can call an earth-centered spherical shell its own private domain if the outer and inner radii are the greatest and smallest distances, respectively, from the Earth to the planet, or at least part of such a shell, truncated symmetrically to the ecliptic to allow the planet to move in latitude.

Further, Ptolemy says [*Planetary Hypotheses* (*PH*) I, ii, 4] that it is most plausible that the spherical shells are snugly nested, "for," he continues, "it is not conceivable that there be in nature a vacuum, or any meaningless and useless thing." Thus, the greatest distance of one planet from the Earth equals the smallest distance of the next.

Ptolemy now accepts that we meet the planets, as we proceed outward from the Earth, in the following sequence: Moon, Mercury, Venus, Sun, Mars, Jupiter, and Saturn, before coming to the fixed stars. Ptolemy mentions this order in *Almagest* IX, 1, but only in passing. This seems quite natural when one considers that his goal with the *Almagest*'s planetary models is to predict *directions* to the planets, and their distances are so great in comparison to the size of the Earth that the place of observation has no effect on the direction of the lines of sight to them. Indeed, if we could measure a difference in direction to a given

Principles, Assumptions, and Construction of the Ptolemaic System 119

planet when viewed from two places on the Earth, we would have a means of finding how far away it is. This is what Ptolemy puts in other words when he says in this passage that the only sure way of finding planetary distances is by determining planetary parallaxes, and none is sensible to the naked eye (I shall return to the problem of finding a parallax). He ascribes the order he accepts to the "older astronomers" and adds that there has been considerable discussion of the matter. Thus, some more recent astronomers have, he says, placed all the planets beyond the sun because no one has ever seen a planet cross the disc of the sun. Ptolemy counters this argument, both in the *Almagest* (IX, 1) and the *Planetary Hypotheses* (I, ii, 2), by saying that such transits would be extremely rare and that they would likely be invisible to the naked eye since a solar eclipse of magnitude equal to a planetary diameter cannot be seen. Thus, he finds it plausible that the sun separates the three planets that can reach opposition (our outer planets) from the two that cannot (our inner planets). Otherwise the order reflects that a slower planet ought to be more distant, and all agree, he says, that the moon is nearer to us, and the fixed stars farther away, than any planet.

The next assumption is completely new with Ptolemy. In the *Almagest* (III, 3) he proves that the size of his epicyclic models does not matter at all, as far as predicting directions to the celestial bodies is concerned, just so long as all proportions and mean motions are correct. In modern times there has been much discussion of whether in the *Almagest* Ptolemy views his models as descriptions of how the planets actually move in space or merely as devices to serve as the basis for calculating planetary positions. That even the well-informed can take part in this discussion shows, of course, that the question is irrelevant to the *Almagest*'s aims. In the *Planetary Hypotheses*, however, there is no doubt on this point, for Ptolemy now assumes that his detailed quantitative models, correctly scaled, really represent the way the planets move around us. (This assumption happens to be very nearly right, as we have seen; it is in his choice of absolute sizes that Ptolemy errs.) This assumption enables Ptolemy to find the ratio of inner to outer radius of each spherical shell. To see how this is done, let us consider his detailed model for a planet (Figure 1) other than Mercury.

The entire model is at first thought of as lying in the plane of the ecliptic through the Earth E—later it is tilted in various ways to represent the planet's movement in latitude. The planet

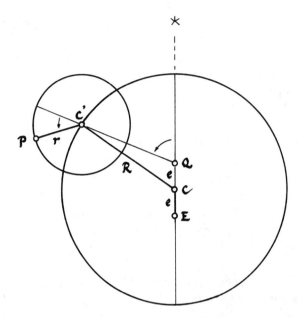

FIGURE 1.

P is carried on the circumference of the epicycle of radius r and center C', which, in turn, moves on a larger circle, the deferent, of center C and radius R. The distance from the Earth E to the center C of the deferent is called the eccentricity, e. The line EC, the apsidal line, remains fixed with respect to the fixed stars. Though it is irrelevant for our present purpose, I should mention that the epicycle's center C' moves uniformly, not around the deferent's center, but as seen from the equant point Q placed on the apsidal line so that $CQ = EC = e$, and that the planet P moves uniformly, not with respect to CC', but to QC'; thus, it is the two angles marked with arrows in Figure 1 that grow uniformly with time. When Ptolemy introduces the equant point, he blithely violates the philosophical doctrine of uniform circular motion as the only permitted component, but it is this bold step that enables him to achieve such excellence so economically.

Such a model is used for Venus as well as for the outer planets. The only difference, except in the necessary parameters, is that in Venus's model the radius of the deferent or rather, more precisely, the line QC' from the equant point to the epicycle's center, points toward the mean sun, while for any outer planet the epicycle's radius $C'P$ to the planet is parallel to that direction.

Principles, Assumptions, and Construction of the Ptolemaic System 121

Ptolemy's model for Mercury is, of necessity, more complicated, and I shall not discuss it here.

Ptolemy measures the dimensions of each model with a unit that is one-sixtieth of the radius of the deferent. The actual size of this unit is irrelevant when we want to find the direction of the line of sight from the Earth to the planet (EP) as Ptolemy has shown. In the *Almagest* he makes no attempt at finding it for any planet, but it is otherwise in the *Planetary Hypotheses*. With the new assumption that a planet really moves according to such a model of correct size, he first determines the ratio of the planet's least-to-greatest distance from the Earth. Indeed, if the planet's spherical shell has inner and outer radii in that same ratio, then it can just accommodate an epicyclic model of the right proportions.

The planet is farthest from the Earth when the center of the epicycle is at the apogee of the deferent and when the planet is at the apogee of the epicycle ("apogee" means the point most distant from the Earth). We then obtain that the greatest distance D of the planet is [see Figure 2(a)]

$$D = R + r + e = 60 + r + e \text{ Ptol. units.}$$

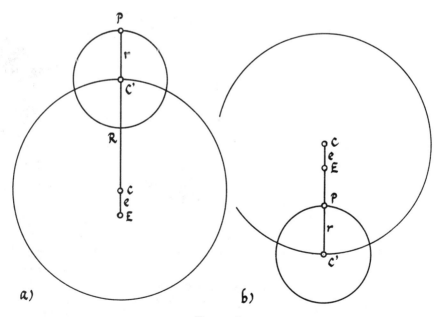

FIGURE 2.

TABLE 1

		$d:D$	d'	D'
Moon	☾	34:65	1	2
Mercury	☿	34:88	2	5
Venus	♀	16:104	5	32
Sun	☉	23:25	32	35
Mars	♂	1:7	35	245
Jupiter	♃	23:37	245	394
Saturn	♄	5:7	394	551

Likewise, we get for closest approach d of the planet [see Figure 2(b)]

$$d = R - r - e = 60 - r - e \text{ Ptol. units.}$$

Thus, the ratio $D:d$ is easily found. Let us consider the model for Venus. Here we have, as everywhere, $R = 60$ and, according to *Almagest* X, 1–5, $r = 43;10$ and $e = 1;15$ (here I follow the now-standard way of writing sexagesimal fractions: 43;10 means 43 10/60, etc.). We then get

$$D = 60 + 43;10 + 1;15 = 104;25$$

and

$$d = 60 - 43;10 - 1;15 = 15;35,$$

so

$$d : D = 15;35 : 104;25.$$

In the *Planetary Hypotheses* Ptolemy rounds this ratio to 16:104. He proceeds likewise for the other planets and obtains the ratios of least-to-greatest distance given in the first column of Table 1. I included the corresponding ratios from the *Almagest* for the Sun and Moon. Ptolemy takes over Hipparchus's solar model without change; it is a simple eccentric circle with an eccentricity of $2\frac{1}{2}$ parts in 60. His complicated lunar model does indeed give rise to the outrageous variation in lunar distance in a ratio of nearly two to one. I cannot go into detail with these matters here but shall merely hint that this awkwardness is the result of Ptolemy's successful attempt at accounting for the lunar inequality now called the "evection," which manifests itself as an enlargement of the moon's possible deviation from mean behavior at first and last quarters. Ptolemy achieves this by pulling the moon's epicycle closer to the earth at quadrature to make it seem larger, but at the price of a much-too-large variation in lunar distance and, one should expect, in the moon's apparent size, but he is silent on this embarrassing point in the *Almagest*.

We are now in a position to construct a scale model of the universe. If we call the inner radius of the lunar sphere $d' = 1$, then the outer radius becomes $D' = 65/34$, or to the nearest integer, 2. The principle of nesting spheres demands that this D' also be d' for Mercury. Applying the ratio of $D:d = 88:34$, we obtain $D' = 5$, again to the nearest integer, which, in turn, must be d' for Venus, and thus we can proceed all the way out through the spheres and obtain the values for d' and D' in Table 1.

If just one of these dimensions could be given in some terrestrial unit of length, the whole universe could then be measured with that unit, and Ptolemy's next step is to introduce the Earth's radius as a yardstick, but, as we shall see, he gets more than he really needs and wants. In the *Planetary Hypotheses* I, ii, 3, he refers to the result obtained in *Almagest* V, 15, where he is concerned with finding the distances to the Moon and Sun in terrestrial radii (t.r.), which he needs for forecasting solar eclipses. His method certainly goes back to Hipparchus, and I present it here in quite ahistorical and much simplified form, in respect of both geometrical argument and parameters, in order to display its essential core.

The argument proceeds from the following two pieces of data:

(i) The apparent diameters of the Sun and Moon are both equal to half a degree, very nearly, and
(ii) the apparent diameter of the Earth's shadow cone, cut at the distance of the Moon, is $2\frac{1}{2}$ times as large as the apparent diameter of the Moon. I shall return to how such estimates could be obtained.

Figure 3 represents—with no pretense whatsoever at being drawn to scale—the Sun and the Earth, with the Earth's shadow cone extending toward the right, and with part of the lunar orbit drawn in. The angles ϕ and ψ are the apparent radii of the Sun and the shadow cone's circular cross-section at the distance of the Moon, respectively, and our data tell us that they are

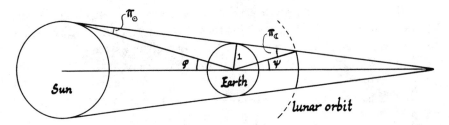

FIGURE 3.

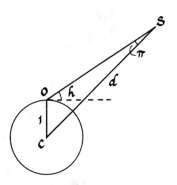

FIGURE 4.

(i) $\phi = \frac{1}{4}°$ and (ii) $\psi = 2\frac{1}{2} \cdot \phi = \frac{5°}{8}$.

The angles marked $\pi_\odot$ and $\pi_\mathbb{C}$ are what we call the horizontal parallaxes of the Sun and Moon, and we get immediately that

$$\pi_\odot + \pi_\mathbb{C} = \phi + \psi = \frac{7°}{8}$$

(ϕ and ψ and the big angle at the Earth's center add up to 180°, but so do $\pi_\odot$, $\pi_\mathbb{C}$, and the same big angle). Thus, if we can determine one of the parallaxes, this relation gives us the other.

To see what is involved in finding the parallax of a celestial object, consider Figure 4. Here C is the center of the earth, O is an observer on the earth's surface whose horizon is indicated by the dotted line, S is the object whose altitude as seen by O is the angle h, and the angle π is what we call its daily parallax. It is, then, the difference between the directions of the lines of sight to S from O and C, respectively, and if we call the terrestrial radius OC 1, and the distance from C to S d, we obtain immediately

$$\frac{\sin \pi}{1} = \frac{\sin(90+h)}{d} = \frac{\cos h}{d}.$$

The parallax vanishes if $h = 90°$, that is, if S is in the observer's zenith, and assumes its largest value when $h = 0$, that is, when S is in the observer's horizon. This last value of π we call the horizontal parallax of S—we encountered it for the sun and moon earlier—and if we measure it in radians, we can replace $\sin \pi$ by π for small values of π and get

$$\pi = \frac{1}{d},$$

where d is measured in earth radii.

Translating the statement about solar and lunar horizontal parallaxes into radians, we get

$$\pi_\odot + \pi_\mathrm{C} = \tfrac{7°}{8} = \tfrac{7}{8} \cdot \tfrac{1}{57.3} \left(= \tfrac{1}{65.5}\right) \text{ radians}$$

or, in terms of solar and lunar distances,

$$\frac{1}{d_\odot} + \frac{1}{d_\mathrm{C}} = \frac{1}{65.5},$$

where $d_\odot$ and d_C are measured in earth radii.

At the end of *Almagest* V, 11, Ptolemy says that Hipparchus found the distance to the moon by assuming the distance to the sun: At one time he took the sun to have a barely observable parallax, and at another he supposed that it had none at all. (Hipparchus's works on these, as on most matters, are lost.) This used to seem a curious procedure, for of the two distances that to the sun is by far the more difficult to measure (in fact, the astronomical unit, as we call it, was not well determined until the end of the 19th century). However, as has now become clear, particularly through the work of Swerdlow and of Toomer, Hipparchus saw that the contribution of lunar parallax to the total of $7/8°$ ($= 1/65.5$ rad.) in the relation above was predominant, for he had come to believe, as was true, that solar parallax was not directly observable with his means. Thus, the relation could serve to give him a good first estimate of the distance to the moon. More precisely, if he assigned to the solar parallax the least observable amount (very likely 7 minutes of arc, as Swerdlow has argued) and then let it vanish, he would obtain a lower and an upper bound for lunar parallax, and hence an upper and a lower bound for lunar distance. Since the two limits would not be very far apart, this approach would give him a good idea of the order of magnitude of the moon's distance from the earth. To illustrate this argument with our crude values above, we would get as the upper limit of lunar parallax, namely with vanishing solar parallax,

$$\pi_\mathrm{C} = \tfrac{7°}{8} = \tfrac{1}{65.5} \text{ rad.}, \quad \text{or} \quad d_\mathrm{C} = 65.5 \text{ t.r.}$$

as the lower limit of lunar distance. This is of the right order of magnitude, though too large by about 10 percent; a solar parallax of $7'$ would increase d_C to 75.5 terrestrial radii. The modern values of lunar and solar mean horizontal parallax are $57'$ and $8.8''$, respectively, to which correspond lunar and solar mean distances of a little over 60 terrestrial radii and about 23,500 terrestrial radii, so the interval from 65.5 to 75.5 terrestrial radii for the lunar distances fails to capture the correct value. This is, of course, due to the crudeness of the two pieces of initial data—the values I assigned to the apparent diameters of the sun and the shadow—and not to any flaw in the method.

As I said, my derivation of the relation between solar and lunar distances from the two pieces of data is very much simplified—this simplification goes back at least to J. L. E. Dreyer. Ptolemy performs his calculations using trigonometry, and so in all probability did Hipparchus before him, and both of them use data different from mine. It seems that Hipparchus found the apparent diameter of the sun (and moon) to be 1/650 of a complete circle ($= 0;31,14°$) by means of his diopter, and he determined the shadow's diameter to be $2\tfrac{1}{2}$ times that of the moon, very likely by measuring with a water clock the time the moon needed to traverse the shadow during a nearly central lunar eclipse.

To digress for a moment, Hipparchus was certainly not the first to find the apparent solar diameter. In his sole surviving work, *On Sizes and Distances of Sun and Moon*, Aristarchus of Samos (third century B.C.) presents the puzzling value of $2°$ (a fifteenth of a sign, as he puts it), while Archimedes (287–212 B.C.) ascribes to him the discovery that it is about 1/720 part of the circle of the zodiac (i.e., $1/2°$). Archimedes tells of this in the preface to his book *The Sand-Reckoner* and proceeds to describe his own determination of that parameter, the earliest explicit account of a scientific measurement that I know of. Archimedes says that he pointed a firmly fixed ruler with a small cylinder standing on it toward the rising sun—one can bear looking at it then—and holding his eye at the ruler's end he moved the cylinder first so it just covered the solar disc and then so the sun could just be seen on either side of it, and marked the two positions of the cylinder. From these two measurements, and even taking into account that the eye does not see from a point but from an area (the pupil) whose size he also determined, Archimedes calculated that the apparent solar diameter is more than 1/200 of and less than 1/164 of a right angle, limits that still hold. It is noteworthy that the result of the earliest measurement is given as an interval, and not as a single number, but it should not be surprising, for we are dealing with Archimedes.

For his derivation of the two diameters, Ptolemy scorns the use of diopter and water clock. Instead he analyzes two lunar eclipse reports from Babylon in the light of his refined lunar theory. During these eclipses one-fourth and one-half of the lunar diameter, respectively, reached into the shadow, and from this he calculates that the lunar diameter must be 0;31,20° and that the shadow's diameter is $2\frac{3}{5}$ times that of the moon. With his version of the above argument, the connection between solar and lunar distance is established.

Already in *Almagest* V, 13, Ptolemy has, however, found the distance to the moon. He does it by determining lunar parallax as the difference between a computed and an observed lunar altitude; this is proper, for his lunar theory represents the moon as seen from the center of the earth. From the parallax he can find the distance from earth to moon in terrestrial radii, as we saw earlier, for one particular configuration of the lunar model, and so he can assign all parts of the model terrestrial scale. I shall omit his observation and calculations and merely mention that he finds the greatest distance D to the moon to be 64;10 t.r.—he abbreviates it to 64 in the *Planetary Hypotheses*. It is with this value that he enters the relation between solar and lunar distances—he maintains that lunar and solar diameters appear equal when the moon is at its farthest from the earth and the sun at its mean distance—and he finds the mean distance to the sun to be 1210 terrestrial radii.

As I said, all we need to measure the universe in terrestrial units is to have just one of the distances in Table 1 so determined, but here Ptolemy presents us with two, which may prove embarrassing: a greatest distance of the Moon of 64 t.r. and a mean solar distance of 1210 t.r., which imply smallest and

TABLE 2. Distances in terrestrial radii (t.r.)

	$d:D$	d	D
☽	34:65	33	64
☿	34:88	64	166
♀	16:104	166	1079
☉	23:25	1160 (1210)	1260
♂	1:7	1260	8820
♃	23:37	8820	14189
♄	5:7	14189	19865

greatest solar distances of 1160 t.r. and 1260 t.r., respectively. Yet we can but try. Proceeding as before, but beginning with $D = 64$ t.r. for the Moon, we find immediately $d = 64$ t.r. for Mercury (see Table 2). We apply the ratio of $D:d$ and find $D = 166$ t.r. for Mercury, which becomes d for Venus. With some trepidation we multiply $d = 166$ t.r. by the ratio 104:16 and obtain $D = 1079$, which miraculously is very close to $d = 1160$ for the Sun. Ptolemy now says (*P.H.* I, ii, 3)

> Since the least distance of the Sun is 1,160 earth radii, as we mentioned, there is a discrepancy between the two distances which we cannot account for; but we were led inescapably to the distances which we set down. So much for the two (planetary) spheres which lie closer to the earth than the others. The remaining spheres cannot lie between the spheres of the Moon and the Sun, for even the sphere of Mars, which is the nearest to the earth of the remaining spheres, and whose ratio of greatest to least is about 7:1, cannot be accommodated between the greatest distance of Venus and the least distance of the Sun. On the other hand it so happens that when we increase the distance to the Moon, we are forced to decrease the distance to the Sun, and *vice versa*. Thus, if we increase the distance to the Moon slightly, the distance to the Sun will be somewhat diminished and it will then correspond to the greatest distance of Venus. [Goldstein's translation]

Ptolemy is right, for the relation

$$\frac{1}{d_\odot} + \frac{1}{d_\mathrm{☾}} = \text{constant}$$

Principles, Assumptions, and Construction of the Ptolemaic System

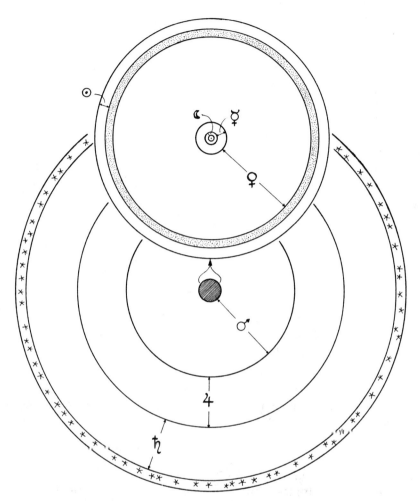

FIGURE 5. The Ptolemaic system (*P.H.* I) drawn approximately to scale with the sub-Martian orbs enlarged tenfold. The earth's diameter is 1/33 of the sublunar orb's.

implies, of course, that if one of the distances is increased, the other must decrease. However, he does nothing to close the slight gap between the spheres of Venus and the Sun but continues building the system outward from the solar sphere, taking D for the Sun to be d for Mars, and so on. Outside Saturn's orb we have the fixed stars at a distance of very nearly 20,000 earth radii, and the system is complete.

In Figure 5 I have drawn the spheres in proper proportion; the

130   3. Ptolemy's Cosmology

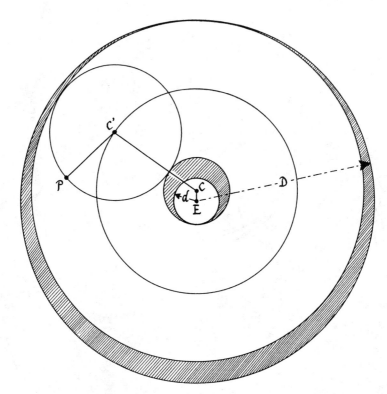

FIGURE 6. *E*: earth; *C*: center of deferent; *C'*: center of epicycle; *P*: planet (Mars); *d*: smallest distance; and *D*: largest distance.

dotted area is the gap between Venus's and the Sun's orbs, and for the sake of clarity I have enlarged the spheres inside Mars's 10 times and moved them upward, out of the way. You are asked to envisage this part compressed to one-tenth of its size and placed into the shaded hole in Mars's sphere.

The cosmological system is now finished and measured in earthly measure—we shall presently transform the dimensions into everyday units. We have seen that it arises as a combination of old and new components, and chief among the new are those of a quantitative character: The spheres are assigned definite sizes and are constructed so that they can accommodate the best quantitative astronomical models of the day. No wonder, then, that the system gained such widespread and lasting popularity. I drew Figure 6 to show how a planetary model fits into the planet's spherical shell. The figure represents the intersection between the sphere and the ecliptic. I chose Mars because its eccentricity is so large that I could afford to draw the model in correct proportion in all its parts. In principle, Mars can be anywhere in the white eccentric annulus, but it can never invade the two shaded lunules—one must remember that C, the center of the deferent, is fixed, whereas C' and P can be anywhere on their respective circles.

Ptolemy now says (*P.H.* I, ii, 4) that the circumference of the earth is 18 myriad (180,000) stades, which agrees with his *Geography*, where he maintains that one degree measured along a meridian (i.e., going due north or south) is 500 stades. From this it follows that one terrestrial radius is 2;52 myriad stades (he actually gives it, Egyptian fashion, as $2 + 1/2 + 1/3 + 1/30$ myriad stades). The *stade* is a Greek unit of length always, as far as we know, equal to 600 feet, but the length of the Greek foot seems to have varied with time and place. Thus, it is impossible to say how good his estimate of the earth's size is, but it is certainly of the right order of magnitude, though very likely quite a bit too small (one degree of latitude is about 364,500 English and American feet, while 500 stades is 300,000 Greek feet). When he adopts this value, he follows Marinos (died circa 110–120), whose work on geography is lost, though a good part of its contents are known, for Ptolemy refers to it extensively. The estimate of 180,000 stades for the earth's circumference, adopted by Marinos, seems to go back at least to Eratosthenes (third century B.C.), according to Cleomedes (fourth century A.D.), but so does another estimate that was very popular before Marinos and Ptolemy and variously given as 250,000 or 252,000 stades (the latter yields very neatly 700 stades per degree of latitude). However, Eratosthenes was certainly not the first to attempt to measure the earth, for his older contemporary and aquaintance Archimedes (ca. 287–212 B.C.) tells that "some" had tried to prove that its circumference is 300,000 stades.

To reach the larger of his two estimates, Eratosthenes proceeded from the following pieces of data:

(i) Alexandria and Syëne (modern Aswan) lie on the same meridian;
(ii) at summer solstice the sun culminates in zenith at Syëne (its rays reach the bottom of a well, it is said), but 1/50 of a full circle south of zenith at Alexandria;
(iii) the distance between the two places is 5000 stades.

The circumference of the earth is then simply 50×5000 stades, or 250,000 stades. The smaller estimate seems to arise in the same fashion, but a locality directly north of Alexandria across the sea plays the role of Syëne.

In his *Geography* Ptolemy shows how one may derive the length of a degree from the latitudes and the difference in longitude of two places of known distance, measured in stades, by means of spherical trigonometry. As he was well aware, there are two difficulties with this procedure. It is fairly easy to establish the latitude of a locality, but it is otherwise when one wants to find the difference in longitude between two places, for it involves comparing their local times. In antiquity it was impossible to carry the time of one location to another—it remained so, within the margin of what is useful, until the end of the 18th century—so one had to rely on lunar eclipses to serve as universal time signals for the comparison. The other difficulty is, of course, with measuring the distance between places that are far apart. In ancient itineraries such distances are most often given by the number of days it took to travel from one to the other, a number much affected by irrelevant circumstances like the nature of intervening terrain.

These two difficulties are also inherent in the older method, for to ascertain that two distant places lie on the same meridian one must show that they have the same longitude (the longitude of Syëne is, in fact, 3° more easterly than that of Alexandria), and it is just as hard to measure a distance in the direction north or south as in any other direction. The only advantage of the older method is that it does not depend on spherical trigonometry, which is just as well, for this subject had not yet been invented at the time of Eratosthenes.

Much has been written in modern times about the quality of the various measurements of the earth and, indeed, if you pick the length of a stade carefully you can show that Eratosthenes's larger value is wonderfully accurate. However, I hope that my short discussion has shown that such efforts are historically irrelevant, for quite apart from the uncertainty about the length of the Greek foot or stade, all underlying data are nothing like modern measurements, but rather rough estimates given in nice round numbers.

To return to the principal matter, we can now simply multiply all the distances in Table 2 by 2;52 to convert them from terrestrial radii into myriad stades, and the universe is measured in terms of a unit that all have experienced by walking over it on

TABLE 3.

	Mean distance in Earth radii	Apparent diameter compared to the Sun's	True diameter compared to the Earth's	Volume compared to the Earth's
Moon	48	$1\frac{1}{3}$	$\frac{1}{4}+\frac{1}{24}$	$\frac{1}{40}$
Mercury	115	$\frac{1}{15}$	$\frac{1}{27}$	$\frac{1}{19,683}$
Venus	$622\frac{1}{2}$	$\frac{1}{10}$	$\frac{1}{4}+\frac{1}{20}$	$\frac{1}{44}$
Sun	1,210	1	$5\frac{1}{2}$	$166\frac{1}{3}$
Mars	5,040	$\frac{1}{20}$	$1\frac{1}{7}$	$1\frac{1}{2}$
Jupiter	11,504	$\frac{1}{12}$	$4\frac{1}{3}+\frac{1}{40}$	$82\frac{1}{4}+\frac{1}{20}$
Saturn	17,026	$\frac{1}{18}$	$4\frac{1}{4}+\frac{1}{20}$	$79\frac{1}{2}$
1. Magn. Stars	20,000	$\frac{1}{20}$	$4\frac{1}{2}+\frac{1}{20}$	$94\frac{1}{6}+\frac{1}{8}$

their own two legs. Ptolemy performs these multiplications in *P.H.* I, ii, 4.

In the final section Ptolemy finds the sizes of the planets, the Sun, and the Moon. Here he has to call on a tradition of estimates of the apparent angular diameters of planets and fixed stars, a matter not discussed in the *Almagest*. The values he chooses to use are given in the second column of Table 3, and they represent the apparent diameters of planets and fixed stars as fractions of the Sun's apparent diameter. He ascribes some of them to Hipparchus and says that he found the others himself, but he does not say how. However, they can surely be nothing but rough estimates, depending more on the brightness of a planet than on the size of its apparent disc, for it was not until after the introduction of the telescope that one could see that planets present a disc and that fixed stars, no matter how bright, remain mere points, no matter how great the magnification.

Ptolemy assumes that these estimates obtain when the celestial bodies are at mean distance, so he next takes the average of d and D in Table 2 and finds the values listed in the first column of Table 3. The too-large value of the apparent lunar diameter does indeed follow from his much-too-small value of the mean distance of the Moon.

Ptolemy now refers us to the *Almagest* (V, 16), where he has shown that the solar radius is $5\frac{1}{2}$ terrestrial radii and, indeed, an angle of 0;31,20°—the Sun's apparent diameter—subtends a chord of very nearly 11 t.r. in a circle of radius 1210 t.r., the

Sun's mean distance. If we now multiply the fractions in the second column of Table 3 by $5\frac{1}{2}$, we would find the true planetary diameters in terrestrial diameters provided that the planets were all placed at the same distance as the Sun, which, of course, they are not. We must then bring each planet to its proper distance or, more precisely, divide by 1210 and multiply by the appropriate mean distance listed in the first column of Table 3 in order to obtain each planet's true diameter in terrestrial diameter. In Mercury's case we must then take the 1/15 from the second column, multiply it by $5\frac{1}{2}$, divide by 1210, and multiply by 115 from the first column. Ptolemy rounds the result to 1/27, which is entered in the third column. [The value 1/28 quoted by Dante from al-Farghani (see p. 115) is, in fact, closer, for our calculation yields 0.0348, in decimals, while $1/27 = 0.0370$ and $1/28 = 0.0357$.] Thus, Ptolemy arrives at the true diameters of the planets compared to the Earth's in Table 3—all manuscripts agree on the value $1/4 + 1/20$ for Venus, but this is an obvious error for the correct $1/4 + 1/30$.

Ptolemy now tacitly assumes that all celestial bodies are spherical like the Earth, and in the last column of Table 3 their volumes are compared to the Earth's. The numbers here are simply the cubes of the numbers in the previous column.

4
Kepler Motion Viewed from Either Focus

In the first chapter we analyzed a Babylonian planetary text and saw that the Late-Babylonian astronomers had shaped arithmetic into a powerful tool for addressing astronomical problems. In contrast, the Greek planetary models, qualitative as well as quantitative, employed geometrical models with moving parts and a knowledge of how to combine velocities. We call such models *cinematical*, for cinematics is the branch of mathematics concerned with motion but without regard to masses and forces. Finally, dynamics—the study of the behavior of a system of masses in terms of the forces that act upon them—was created in its useful form by Isaac Newton (1642–1727). In his *Philosophiae naturalis principia mathematica (Mathematical Principles of Natural Philosophy)*, usually called the *Principia* (1687), he formulated the basic rules of dynamics and successfully applied them to a variety of physical problems, among them that posed by the solar system.

The last major chapter in the story of the cinematical approach to planetary astronomy was written by Johannes Kepler (1571–1630), though he himself would very likely have objected vehemently to this characterization of his work, as we shall see. By degrees he formulated the laws that still bear his name, among them his discovery that each planet, including earth, travels in a planar, elliptic orbit with the sun in one focus and such that the area swept out by the line joining sun and planet grows uniformly with time. More specifically, as Figure 1 illustrates, he found that the planet P travels in an ellipse whose major axis remains very nearly fixed relative to the fixed stars and with the sun S at one focus. The far endpoint of the major axis from S is called the aphelion A ("aphelion" means the point farthest from the sun), and the planet moves so that the shaded

4. Kepler Motion Viewed from Either Focus

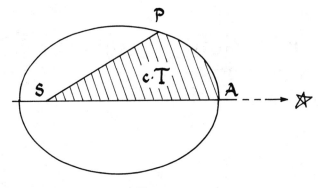

FIGURE 1.

area ASP is $c \cdot T$, where c is an appropriate constant and T the time elapsed since the planet was last at aphelion.

In the present chapter, I shall try to make clear what is involved in finding a planetary position when planet and earth are known to obey Kepler's rules; then I shall briefly outline how Kepler himself solved this problem; finally, I shall show that Ptolemy's equant model owes its success to a certain property of Keplerian motion—that if you observe a planet, not from the sun, but from the other and empty focus of its orbit (a much more comfortable place to be), then the planet will seem to travel very nearly uniformly around you.

Properties of Ellipses

Before I turn to these things, it might be well if I briefly review some of the results and techniques we shall need in the following, particularly such that concern ellipses. In current texts on analytic geometry the ellipse is commonly introduced as the locus of points satisfying a certain geometrical condition. Let there be given two points, F_1 and F_2, and a distance $2a > F_1F_2$; then the ellipse is the locus of the points P the sum of whose distances from F_1 and F_2 is always $2a$, namely,

$$F_1P + F_2P = 2a.$$

If we now introduce the eccentricity e as the ratio between F_1F_2 and $2a$, and a coordinate system as in Figure 2, then we can derive the equation of the ellipse with the standard techniques of analytic geometry—we use the distance formula—and after a

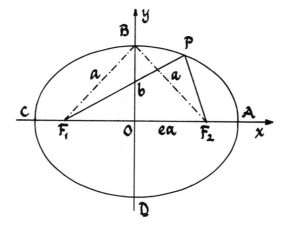

FIGURE 2.

reduction we get

$$\frac{x^2}{a^2} + \frac{y^2}{(1-e^2)a^2} = 1,$$

or, setting $(1 - e^2) \cdot a^2 = b^2$,

$$\frac{x^2}{a^2} + \frac{y^2}{b^2} = 1.$$

The points A, B, C, D in Figure 2 are called the vertices of the ellipse, F_1 and F_2 are its foci, O is its center, the distance $OA = a$ its semi-major axis, and the distance $OB = a\sqrt{1-e^2} = b$ its semi-minor axis.

When one reads older texts it is useful to know that *latus rectum* is the term used for the chord through a focus and perpendicular to the major axis. For its length we get

$$l = \frac{2b^2}{a} = 2a(1-e^2).$$

If we solve the equation of the ellipse for y, we obtain

$$y = \pm \frac{b}{a}\sqrt{a^2 - x^2},$$

where the signs refer to upper and lower halves of the curve, respectively. If P_1 of coordinates (x, y_1) is a point of the ellipse's circumscribed circle, which has center O and radius a, we find

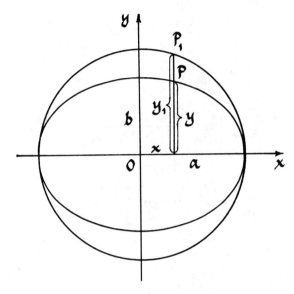

FIGURE 3.

$$x^2 + y_1^2 = a^2,$$

or again

$$y_1 = \pm\sqrt{a^2 - x^2}.$$

If P and P_1 are two points, the former on the ellipse and the latter on the circle, with the same abscissa and on the same side of the major axis (see Figure 3), we get for their ordinates, from the preceding equations

$$y = \frac{b}{a} y_1.$$

We may then think of the ellipse as the curve we obtain when we compress the circumscribed circle in a constant ratio $(b:a)$ in a direction perpendicular to the major axis. Thus, if you think of the areas above the major axis and below the upper halves of ellipse and circle as made up of a great many, very narrow vertical strips, and likewise for the lower halves, it should become clear that the areas of ellipse and circle have the same ratio as the ordinates. The circle's area is πa^2, so the area enclosed by the ellipse is

$$\frac{b}{a} \cdot \pi a^2 = \pi a b.$$

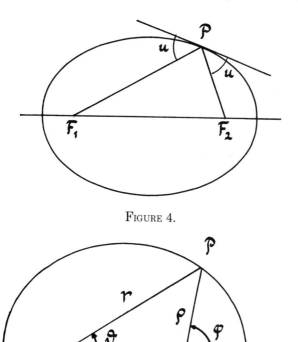

FIGURE 4.

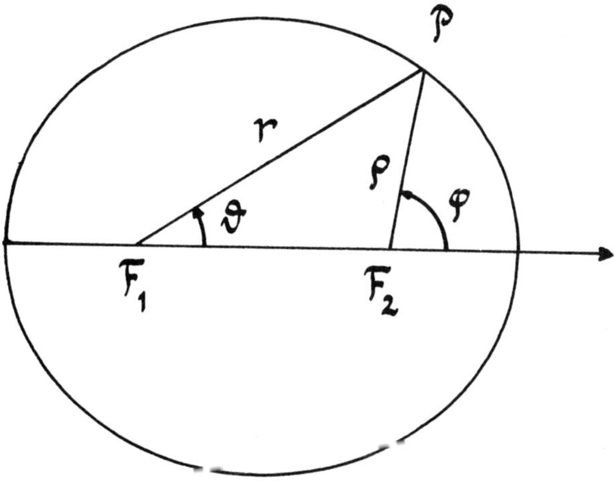

FIGURE 5.

One more property of the ellipse will be of use in the following. If P is any point on the ellipse, then the focal rays to P, F_1P and F_2P, will form equal angles u with the tangent to the ellipse at P; see Figure 4. This implies that the ellipse will reflect rays emitted from one focus so that they pass through the other—hence the name "focus" for the points F_1 and F_2.

Finally, the equation of an ellipse in polar coordinates takes on a particularly pleasing form if the pole is placed at a focus. With the pole at F_1 and the polar axis along the major axis as in Figure 5, the equation of the ellipse is

$$r = \frac{\frac{l}{2}}{1 - e\cos\theta} = \frac{a(1-e^2)}{1 - e\cos\theta},$$

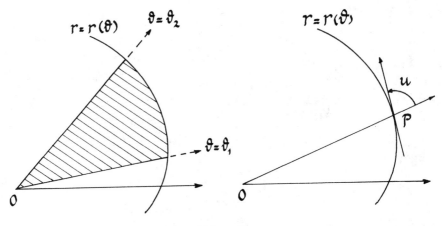

FIGURE 6.

and with the pole at F_2 and the polar axis as before, it is

$$\rho = \frac{\frac{l}{2}}{1+e\cos\varphi} = \frac{a(1-e^2)}{1+e\cos\varphi}.$$

These formulas can be derived, or verified, with the standard techniques of analytic geometry.

Two formulas that have to do with curves in a polar coordinate system will be employed at the end of this chapter, and I may as well include them in this swift review. If we are given a curve whose equation in polar coordinates is $r = r(\theta)$, then the area enclosed by the curve and the rays $\theta = \theta_1$ and $\theta = \theta_2$ [see Figure 6(a)] is

$$\int_{\theta_1}^{\theta_2} \frac{1}{2} r^2 \, d\theta.$$

Furthermore, if P is a point on the curve, and u the angle from the polar ray through P to the forward half of the tangent at P [see Figure 6(b)], we have

$$\tan u = r \cdot \frac{d\theta}{dr}$$

or, where appropriate, for example, when $dr/d\theta = 0$,

$$\cot u = \frac{1}{r} \cdot \frac{dr}{d\theta}.$$

All these things were known in antiquity except those that have to do directly with rectangular or polar coordinate systems. In fact, in matters concerning the geometry of the ellipse, Kepler refers to Archimedes (ca. 287–212 B.C.) and Apollonius (ca. 200 B.C.) as his authorities. The ellipse is one member of a family of curves that the Greeks called the conic sections, for these curves were first thought of as generated when you cut a conical surface with a plane.

Before Apollonius the surface was always assumed to be that of a right circular cone, and the cutting plane had to be perpendicular to a generator—various attempts, none quite convincing, have been made at explaining this curious convention. With this old definition it is then the angle at the vertex of the cone that determines the kind of conic we get: When this angle is acute, we obtain an ellipse; when right, a parabola; and when obtuse, a hyperbola. The earlier Greeks consequently called the three kinds by the names "acute-angled," "right-angled," or "obtuse-angled" conic sections, and Archimedes still used this terminology in the books where he talked of such curves.

Apollonius's great work on conic sections was in eight books of which the first four survive in the Greek, the next three only in Arabic, while the last is lost. Here he began with a new definition of conic section and introduced the names for them that we still use. For him the conical surface could be that of any circular cone, right or skew, and he considered any plane cut in such a surface as one of his conic sections. He showed, incidentally, that, despite the more relaxed definition, his family of curves is identical to that obtained under the older convention.

I should emphasize that for Kepler the works of these and other ancient authors were not objects of historical study, but treasured sources of technique and knowledge.

Kepler's Equation

We are now prepared to address the problem that has been central in planetary theory since the work of Ptolemy: Given the time, find the direction from us to the planet in question. We saw how Ptolemy solved it, but now we assume that the planets and the earth obey Kepler's rules, which, of course, they do, except for minor perturbations caused by mutual attraction between the planets.

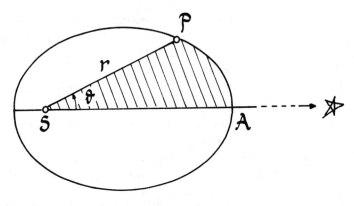

FIGURE 7.

The first step toward that goal consists in finding the position of a planet in its orbit at a given time. Figure 7 represents the planet P on its ellipse with the sun S at one focus and the apsidal line SA always pointing to very nearly the same place among the fixed stars. Before we begin our calculation, we must be provided with information about the orbit: the direction of its apsidal line and its size and shape or, in other words, its semi-major axis a and its eccentricity e. I ought to include among these elements two more indicating how much, and around which line through S, the plane of the orbit is tilted against the plane of the ecliptic, which is that of the earth's orbit. However, for the five classical planets the angles of inclination are rather small, so for the sake of simplicity we may assume that our planet's orbit lies in the ecliptic without much consequence for the planet's longitude. Further we must know the planet's period p, usually given in days, that is, the constant number of days it takes the planet to travel once around its orbit (e.g., from the aphelion A and back again) as well as time $t(A)$ when it last was at A [all possible values of $t(A)$ are separated by whole multiples of p].

I cannot afford here to describe how these elements were first derived from observations and later refined by degrees, for this story is as complicated as it is worth telling. For our present purpose it suffices to take them as given.

If we are then presented with an arbitrary time t and wish to find where in its orbit the planet is at that moment, we must first find how long a time interval T has elapsed since the planet last was at aphelion and obtain

$$T = t - t(A).$$

The planet moves so that the shaded area in Figure 7 is proportional to T, that is,

$$\text{area } ASP = c \cdot T,$$

where c is a constant we can find from the provided elements. Indeed, when T equals p, the period of the planet's motion, then the radius vector SP will have swept out the entire area of the ellipse since the planet last was at aphelion, and we get

$$\text{area of ellipse} = \pi ab = c \cdot p,$$

which yields

$$c = \frac{\pi ab}{p},$$

and so

$$\text{area } ASP = \frac{\pi ab}{p} \cdot T.$$

We now have the area expressed in terms of time and orbital elements, and we will endeavour to find the planet's position from this or, more precisely, we will seek the radius vector $r = SP$ and the heliocentric elongation θ from the aphelion A (see Figure 7).

To that end we introduce an intermediary quantity E, already called the eccentric anomaly by Kepler, in the following fashion (see Figure 8). First we draw the orbit's circumscribed circle, of center O and radius a. From the planet P we drop the perpendicular PP_0 to the apsidal line and extend it the other way until it intersects the circumscribed circle at P_1. It is the angle AOP_1 that we call the eccentric anomaly E, and we shall now express the uniformly growing area in terms of E and so establish a connection between the time T and E.

Here we notice first that the area bounded by the apsidal line, the circumscribed circle, and the line SP_1 consist of two simple parts:

$$\text{area } ASP_1 = \text{sector } AOP_1 + \text{triangle } OSP_1.$$

If we measure E in radians, we have

$$\text{sector } AOP_1 = \tfrac{1}{2} a^2 E$$

(note that the area of the sector is proportional to E and that to $E = 2\pi$ corresponds the area of the whole circle, πa^2). Triangle

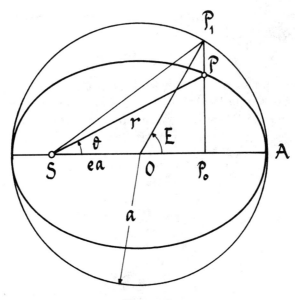

FIGURE 8.

OSP_1 has as its base $OS = ea$ and as its altitude P_0P_1, which is

$$P_0P_1 = OP_1 \cdot \sin E = a \sin E,$$

so we have

$$\triangle OSP_1 = \tfrac{1}{2} ea^2 \sin E.$$

Thus, we get

$$\text{area } ASP_1 = \tfrac{1}{2}a^2 E + \tfrac{1}{2}ea^2 \sin E = \tfrac{1}{2}a^2(E + e \sin E).$$

We recall that we obtain the ellipse if we compress its circumscribed circle in the constant ratio $b:a$ in the direction perpendicular to its major axis, and note that if we reduce the ordinates of the points on the line segment SP_1 in the same ratio, we obtain the ordinates of the points of SP (we think of the major axis as the abscissa axis). It should then be clear that the areas ASP and ASP_1 also have the ratio $b:a$, and we have

$$\text{area } ASP = \frac{b}{a} \cdot \text{area } ASP_1 = \frac{1}{2}ab(E + e \sin E).$$

Combining this with our earlier result that

$$\text{area } ASP = \frac{\pi ab}{p} \cdot T,$$

we obtain
$$E + e \sin E = 2\pi \cdot \frac{T}{p}.$$

It is customary to denote the right-hand side by M. We note that M is a quantity that grows uniformly with time; when T goes from 0 to p, M increases from 0 to 2π, as do E and θ, and M can therefore be thought of as the measure in radians of an angle whose right side is, say, SA and whose left side performs one revolution uniformly while the planet travels from aphelion and back again. We call M the *mean anomaly*. With this notation we have

$$\boxed{E + e \sin E = M.}$$

I should remark that M is not on Figure 8. When M is 0 or π, M and E are equal; otherwise, $M > E$ for the upper half of the orbit in Figure 8 where $e \sin E$ is positive, and $M < E$ for the lower half. If the eccentricity e is zero, so the orbit is a circle, we always have $E = M$ and so $\theta = M$. Uniform circular motion is thus a possible, though unlikely, Keplerian motion.

The boxed formula is Kepler's equation, and it allows us to determine E when the time, and so M, is given. Finding E from this equation is not a trivial matter, and I shall postpone a discussion of how it can be attempted, but let us assume for the moment that it has been done so that we now know E.

Once we have the eccentric anomaly E, our difficulties are over, for we can then easily find the position of the planet P in its orbit or, more precisely, its distance r from the sun and its heliocentric elongation from the aphelion, θ (see again Figure 8). We have

$$P_0 P_1 = a \sin E,$$

so

$$P_0 P = \frac{b}{a} \cdot P_0 P_1 = b \sin E = \sqrt{1 - e^2} \cdot a \sin E.$$

Furthermore,

$$O P_0 = a \cos E,$$

so
$$SP_0 = ea + a\cos E = a(e + \cos E).$$

Since $\triangle SP_0 P$ is a right triangle, we have
$$r^2 = SP^2 = SP_0^2 + P_0 P^2 = a^2(e + \cos E)^2 + (1 - e^2)a^2 \sin^2 E,$$
from which we get, after a slight reduction,
$$r^2 = a^2(1 + 2e \cos E + e^2 \cos^2 E) = a^2(1 + e \cos E)^2,$$
so
$$r = a(1 + e \cos E).$$

As to θ, we have from $\triangle SP_0 P$
$$\cos \theta = \frac{SP_0}{SP} = \frac{a(e + \cos E)}{r} = \frac{e + \cos E}{1 + e \cos E},$$
from which θ is readily found.

Thus, a knowledge of E allows us to determine r and θ through the simple formulas

$$\boxed{r = a(1 + e \cos E) \quad \text{and} \quad \theta = \arccos \frac{e + \cos E}{1 + e \cos E}.}$$

The second formula implies the prettier formula
$$\tan \frac{\theta}{2} = \sqrt{\frac{1 - e}{1 + e}} \cdot \tan \frac{E}{2}.$$

For the sake of completeness, I repeat here the formula from page 139:
$$r = \frac{a(1 - e^2)}{1 - e \cos \theta}.$$

Resolution of Kepler's Equation

We do not meet Kepler's area law as we know it until his *Epitome of Copernican Astronomy*, published in four parts from 1617 to 1622, but Kepler's equation appears already in *Astronomia Nova* of 1609. Here he talks not about the area of a part of the elliptic orbit, but of the area inside the circumscribed circle—

the eccentric circle, as he calls it—which we call area ASP_1 in Figure 8. He shows that it provides a measure of "the sum of all the radii" from the sun to the corresponding part of the ellipse. Indeed, the left side of our form of Kepler's equation

$$E + e \sin E = M,$$

when multiplied by $\frac{1}{2}a^2$ is the area ASP_1, as we saw in the course of its derivation. Here is what Kepler has to say about his equation (*Astronomia Nova*, Chapter 60):

If the mean anomaly (our M) is given, then there exists no geometrical method for deriving either the corrected anomaly (our θ) or the eccentric anomaly (our E). For the mean anomaly is composed of two areas, one sector (area AOP_1 in Figure 8) and one triangle (OSP_1). While the former is measured by the arc on the eccentric circle (the orbit's circumscribed circle), one obtains the latter by multiplying the sine of this arc with the area of the greatest triangle ($\frac{1}{2}ea^2$) and cutting off the last places. There are infinitely many ratios between an arc and the corresponding sine. If then the sum of the two be given, it is therefore impossible to say how large the arc may be, and how large its sine, that correspond to this sum; i.e. unless we first find out how large the area is that belongs to a given arc, i.e. unless we construct tables of $(E + e \sin E)$ and use them backwards.

That is my own way of thinking. The more wanting in geometrical beauty it may appear to be, the more strongly do I urge the geometers to solve for me the following problem:

Given the area of a part of a semicircle, and given a point of the diameter, to find an arc, and an angle at the point, so that the sides of the angle and the arc enclose the given area. Or: to cut the area of a semicircle in a given ratio from an arbitrary given point on the diameter.

It is enough for me to believe that a solution *a priori* is impossible because the arc and the sine are heterogeneous in nature. But whoever proves me wrong and shows the way (to a solution) will seem to me as great as Apollonius.

Terseness was not a characteristic of Kepler's prose style, but he is right when he says that a solution is impossible if by that he means that E cannot be explicitly written in terms of the kind of functions of e and M that he knew, for example, trigonometric functions and their inverses. The equation is what we call transcendental, for one can show that its solution cannot be expressed as a finite combination of elementary functions of e and M. [I here insert that by an "elementary function" we still mean one that L. Euler (1707–83) dealt with in his *Introductio in*

Analysin Infinitorum (1748), the ancestor of all textbooks on calculus; they are essentially exponential and trigonometric functions, their inverses, and algebraic functions.] That a solution exists is obvious, for the function

$$y = x + e \sin x, \qquad 0 \le e < 1,$$

has a positive derivative $(1 + e \cos x \ge 1 - e > 0)$, so it is strictly increasing and therefore has an inverse.

Kepler suggests that though it is hard to find E when M is given, it is easy to find M when we know E (and e). Thus, one can readily make a table of $M = E + e \sin E$ for a dense set of values of E and use it backward. As we shall see, this is in effect what he does. The disadvantage of this method is that we must make a separate table for each value of e, that is, for each planet.

A considerable literature grew up on the solution of Kepler's equation—already Newton gave much attention to this problem—but if all you want is a specific, numerical solution, it can be obtained swiftly and painlessly, particularly if you have at hand a little calculator with trigonometric functions built in. The following iteration method works well.

First a remark on units: We have so far assumed that E and M in Kepler's equation were measured in radians, but in astronomy one prefers to work with degrees. We therefore multiply Kepler's equation by $180/\pi$ and get

$$\frac{180}{\pi} \cdot E + \frac{180}{\pi} \cdot e \sin E = \frac{180}{\pi} \cdot M$$

or, if we now let E stand for the measure of that angle in degrees, and likewise for M,

$$E + \frac{180}{\pi} \cdot e \sin E = M.$$

It might have been more elegant and, more importantly, clearer if I had introduced a particular typeface for the measure of the angles E and M when expressed in degrees rather than in radians, but I adhere to astronomical custom. However, I shall set

$$\frac{180}{\pi} \cdot e = \varepsilon$$

so that Kepler's equation in the form

TABLE 1

e	0.0933	0.2056
M	60°	60°
E_1	55.370491	49.798210
E_2	55.601327	51.002710
E_3	55.589123	50.844860
E_4	55.589767	50.865317
E_5	55.589733	50.862662
E_6	55.589735	50.863007
E_7	55.589735	50.862962
E_8		50.862968
E_9		50.862967
E_{10}		50.862967

$$E + \varepsilon \sin E = M$$

reminds us that E and M are measured in degrees (sin E is, of course, the same no matter what unit E is measured in).

We now rewrite the equation as

$$E = M - \varepsilon \sin E,$$

and from a given value of M we compute a sequence of what turns out to be ever-better approximations of E, beginning with $E_0 = M$:

$$E_1 = M - \varepsilon \sin M,$$
$$E_2 = M - \varepsilon \sin E_1,$$
$$E_3 = M - \varepsilon \sin E_2,$$
$$\vdots$$

If, or rather when, an E_{n+1} turns out to be equal to E_n (to the desired number of decimal places) we have a solution of the equation. One can prove in general that this process converges for $0 \leq e < 1$, but this is not relevant for each specific instance, for the method is self-checking. In Table 1 I listed two sequences of E_n's, one for Mars ($e = 0.0933$) and one for Mercury ($e = 0.2056$), to show how quickly it works (the fractional parts of E are written decimally).

To show how Kepler himself attacked his problem in practice, I reproduced a piece of the *Rudolphine Tables* (1627) in Table 2. It consists of the top and bottom of the first of three pages devoted to the anomalies of Mars (Kepler's page is too long to be

TABLE 2

62   *Tabularum Rudolphi*

Tabula Æquationum MARTIS.

Anomalia Eccentri, *Cum aquationis parte phy.*	Intercolumnium, *Cum Logarithmo.*	Anomalia coæquata.	Intervallū *Cum Logarithmo*		Anomalia Eccentri, *Cum aquatio nis parte phy.*	Intercolumnium, *Cum Logarithmo.*	Anomalia coæquata.	Intervallū *Cum Logarithmo*	
0 ′ ″ 0. 0. 0	Par. ′ ″	Gr. ′ ″ 0. 0. 0	166465 50962	0	30 ′ ″ 2.39.14	15960 0.51. 9	Gr. ′ ″ 27.26.37	164572 49818	12
1 0. 5.34	18130 0.50. 3	0.54.41	166462 50960	0	31 2.44. 2	15810 0.51.13	28.21.57	164447 49742	12
2 0.11. 7	18130 0.50. 3	1.49.22	166456 50957	0	32 2.48.48	15650 0.51.18	29.17.19	164319 49664	12
3 0.16.40	18120 0.50. 3	2.44. 3	166446 50950	1	33 2.53.31	15490 0.51.23	30.12.44	164187 49584	13
4 0.22.13	18110 0.50. 3	3.38.44	166431 50942	1	34 2.58.10	15320 0.51.29	31. 8.11	164051 49501	13
5 0.27.46	18090 0.50. 4	4.33.25	166412 50930	2	35 3. 2.46	15150 0.51.34	32. 3.41	163912 49416	13
6 0.33.18	18070 0.50. 5	5.28. 7	166388 50916	2	36 3. 7.18	14970 0.51.39	32.59.14	163769 49329	14
7 0.38.50	18040 0.50. 6	6.22.49	166360 50899	3	37 3.11.46	14790 0.51.45	33.54.50	163623 49240	14
8 0.44.21	18010 0.50. 7	7.17.32	166328 50879	3	38 3.16.10	14600 0.51.51	34.50.29	163474 49149	14
...	17970	...	166...		...	...	...	...	
21 1.54. 8	17090 0.50.34	19.10.21	165396 50396	8	51 4. 7.31	11800 0.53.17	46.59. 9	47767	19
22 1.59.18	16980 0.50.38	20. 5.21	165437 50342	9	52 4.10.59	11650 0.53.24	47.55.38	161039 47648	19
23 2. 4.25	16870 0.50.41	21. 0.23	165343 50285	9	53 4.14.22	11410 0.53.32	48.52.11	160844 47527	19
24 2. 9.30	16760 0.50.44	21.55.27	165245 50226	10	54 4.17.40	11180 0.53.39	49.48.48	160646 47404	20
25 2.14.33	16640 0.50.48	22.50.33	165143 50164	10	55 4.20.53	10940 0.53.47	50.45.30	160446 47279	20
26 2.19.34	16520 0.50.52	23.45.41	165036 50100	10	56 4.24. 2	10700 0.53.55	51.42.16	160244 47151	20
27 2.24.33	16390 0.50.56	24.40.52	164926 50033	11	57 4.27. 6	10450 0.54. 3	52.39. 6	160039 47024	20
28 2.29.29	16250 0.51. 0	25.36. 5	164812 49964	11	58 4.30. 6	10200 0.54.11	53.36. 0	159830 46894	21
29 2.34.23	16110 0.51. 4	26.31.20	164694 49892	11	59 4.33. 1	9940 0.54.19	54.32.58	159621 46763	21
30 2.39.14	15960 0.51. 9	27.26.37	164572 49818		60 4.35.50	9690 0.54.27	55.30. 0	159409 46630	

TABLE 3

E_n	$\left\| \ln \dfrac{\Delta\theta_n}{\Delta M_n} \right\| \times 10^5$	θ_n	$r_n \times 10^5$
$\varepsilon \sin E_n$	$\dfrac{\Delta\theta_n}{\Delta M_n}$ (sexag.)		$\|\ln r_n\| \times 10^5$

$\varepsilon = e \cdot \dfrac{180}{\pi}$, $e = 0.09265$, $a = 1.52350$ a.u.;

E: eccentric anomaly;
M: mean anomaly, from aphelion; $M_n = E_n + \varepsilon \sin E_n$;
θ: heliocentric longitude, from aphelion;
r: distance from sun in astron units;
$\Delta M_n = M_n - M_{n-1}$; $\Delta\theta_n = \theta_n - \theta_{n-1}$.

adapted to the present format and remain legible). There is an entirely analogous set of three pages for each of the other planets and for the sun—we shall presently see why this last set is for the sun and not the earth.

The page is divided into two columns, each of which consists of 31 rows, except for headings. Such a horizontal row is built up of four little boxes, and I presented the structure of such a row in schematic form in Table 3. The upper number in the first or leftmost box represents the eccentric anomaly E in degrees, and E is the independent variable of the tables in the sense that Kepler chose values for E first and then derived the rest from them. There is a row corresponding to every whole number of degrees of E from 0° to 180°. The index n refers now to the number of the row of boxes; it has nothing to do with the indexes in Table 1.

The third box gives in degrees, minutes, and seconds what Kepler calls the corrected anomaly, our θ, calculated from his version of our formula

$$\cos\theta = \frac{e + \cos E}{1 + e \cos E},$$

where his value of the eccentricity for Mars is $e = 0.09265$.

The upper number in the fourth box represents the distance from sun to Mars, our r, measured in astronomical units and calculated from the equivalent of our formula

$$r = a(1 + e \cos E),$$

with the mean distance $a = 1.52350$ a.u. Kepler gives these numbers to six places, but without decimal points, so that they in fact are $r \cdot 10^5$.

As already remarked, it may be difficult to solve Kepler's equation
$$E + \varepsilon \sin E = M$$
for E when the time, and so M, is given, but it is easy to find M when E is given, and that is, in fact, what Kepler does, keeping E as his independent variable for a short while longer. Indeed, the lower number in the first box is simply $\varepsilon \sin E$, expressed in degrees, minutes, and seconds and computed from the value of E just above it (Kepler calls the quantity $\varepsilon \sin E$ "the physical part of the equation" or correction). The sum of the two numbers in the first box is, then, the value of M to which correspond the values of θ and r in the same row. In other words, Kepler has very simply succeeded in turning the table of equally spaced values of E and the values of θ and r derived from them into a table correlating *unequally*, but still closely, spaced values of M with their corresponding values of θ and r.

These planetary tables are preceded by a set of mean motion tables that enable you to find the mean anomaly M corresponding to any given time t for each planet. I shall describe the simple structure of these tables later. If the value of M you obtain in this way happens to be one of the 181 sums of the two numbers in the first boxes, you are provided directly with the corresponding θ and r. If not—and this is, of course, much more likely—you must find θ and r by linear interpolation, and some of the remaining numbers in the boxes are there to help in this process.

Example

Given $M = 60°$, find θ and r from Table 2.

The nearest values of M in Table 2, and their corresponding values of θ and r, are
$$M = 55° + 4;20,53° = 59;20,53° \approx \theta = 50;45,30° \approx r = 1.60446$$
and
$$M = 56° + 4;24,2° = 60;24,2° \approx \theta = 51;42,16 \approx r = 1.60244.$$
The lower number in the second box belonging to $E = 56°$ is the ratio of the difference between the two θ-values and the two M-values, expressed sexagesimally, so we learn that the incre-

ment in θ corresponding to an increment in M of $1°$ in this interval is $0;53,55°$. Our value $M = 60°$ is $0;24,2°$ smaller than $60;24,2°$, so the associated value of θ is found thus:

from Table 2 : $\theta = 51;42,16$

$$\underline{0;53,55 \cdot 0;24,2 = \ 0;21,36}$$

$$\theta = 51;20,40°.$$

so for $M = 60°$ we get by subtraction

Likewise we find, but now by direct interpolation without the benefit of a provided interpolation coefficient, that to $M = 60°$ corresponds $r = 1.60320$ a.u.

To test the reliability of Kepler's tables, I computed θ and r directly from $M = 60°$, but using his values of eccentricity and mean distance, $e = 0.09265$ and $a = 1.52350$, for Mars. Solving Kepler's equation, I found the eccentric anomaly $E = 55.6189322°$ and hence

$$\theta = \arccos \frac{e + \cos E}{1 + e \cos E} = 51.342568° = 51;20,33,14°$$

compared to his $\theta = 51;20,40°$, and

$$r = a(1 + e \cos E) = 1.60321 \text{ a.u.}$$

compared to his $r = 1.60320$ a.u.

The rest of the numbers in Kepler's tables are, as indicated in his headings and in Table 3, logarithms of the interpolation coefficients and of r provided to be used in further calculations in combination with his various logarithmic tables in the Rudolphine Tables. Kepler's logarithms were close to, but not identical with, our natural logarithms, but I cannot go into these matters here.

Remark

Those already familiar with planetary astronomy will have noticed that the formulas I derived and used look slightly different from the ones they learned from modern textbooks. The reason is that for obvious reasons I adhered to the older convention, introduced by Kepler and followed by Newton and many others, of referring the motions to the aphelion instead of the perihelion, as is the current practice.

154   4. Kepler Motion Viewed from Either Focus

Indeed, if M, E, and θ are counted from the perihelion (the point on the orbit nearest the sun), Kepler's equation will now be

$$E - e \sin E = M,$$

and we further have

$$\cos \theta = \frac{\cos E - e}{1 - e \cos E}$$

and

$$r = a(1 - e \cos E),$$

as is readily checked.

Heliocentric and Geocentric Longitudes

Since the work of Ptolemy, celestial longitudes have been counted along the ecliptic from the point occupied by the sun at spring or vernal equinox. This point marks the beginning of the zodiacal sign Aries on the celestial sphere and is consequently marked ♈0°. It is, then, the direction from the earth to the sun at vernal equinox that provides the basis from which geocentric longitudes are counted.

We have hitherto indicated the direction from the sun to a planet relative to the planet's apsidal line, but no two of these lines coincide. When we now introduce heliocentric longitude to indicate directions from the sun in the plane of the ecliptic (that of the earth's orbit), we reckon it from the same basic direction as geocentric longitude (note that two parallel directed lines have either the same or opposite directions). This convention assures that the geocentric longitude and the heliocentric longitude of a fixed star are equal, for the star is, to all practical purposes, infinitely far away so that the lines of sight to it from earth and sun are parallel.

Figure 9 represents the earth's orbit with the sun S in one focus—here, as elsewhere, I have greatly exaggerated the eccentricity and have let the apsidal line point in a direction that has no particular relation to the correct one. E_v and E_a are the positions of the earth at vernal equinox and autumnal equinox, respectively, and E_v, S, and E_a lie on a straight line as indicated. The direction to ♈0° is, as said, the direction from E_v to S, so the

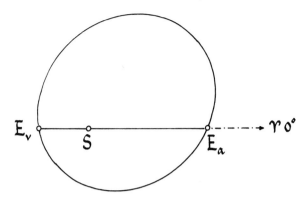

FIGURE 9.

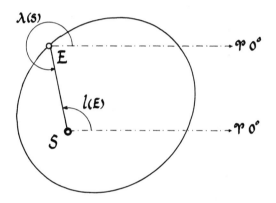

FIGURE 10.

basic direction, the zero direction, or what you will, from which we measure heliocentric longitudes is that from S toward E_a, the place of the earth at autumnal equinox.

In Figure 10 $l(E)$ is the heliocentric longitude of the earth and $\lambda(S)$ the geocentric longitude of the sun. As the figure shows, they differ by 180°. This simple relation is crucial for what follows. Both $\lambda(S)$ and $l(E)$ increase by 360° in one year in the counterclockwise sense if the ecliptic—the plane of the paper—is viewed from the north.

Figure 11 indicates what we need to do when we wish to find the heliocentric longitude $l(P)$ of a planet P. We must be provided with the heliocentric longitude $l(A)$ of the aphelion A and increase it by the corrected anomaly θ. Kepler gives us the means of finding $l(A)$ in the mean motion tables that precede

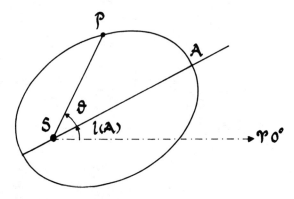

FIGURE 11.

the anomaly tables for each planet. We obtain θ either via the eccentric anomaly E from Kepler's equation or by interpolation in his tables and have

$$l(P) = l(A) + \theta.$$

To enter Kepler's mean motion tables we must be provided with a value of the time, t, given in terms of year, month, day, and hour. He reckons his years according to the Julian Christian calendar in which every fourth year is a leap year so the average length of the year becomes precisely 365 1/4 days. He gives chronological tables for converting dates in other calendars—Jewish, Islamic, and Gregorian—into Julian dates.

The Gregorian calendar, the one we now follow, was introduced by Pope Gregor XIII in 1582 by a papal bull that decreed that in that year October 4 should be followed by October 15. Its aim was to bring vernal equinox back to March 21 and keep it always near this date on which it had fallen at the time of the Council of Nicea in 325 when rules for calculating Easter had been laid down (Easter Sunday should be the first Sunday after the full moon that happens on, or next after, the day of vernal equinox; to calculate its date in advance is a problem of some nicety). The Julian year of 365.25 days is slightly longer than the tropical (seasonal) year of 365.2422 days, so vernal equinox had slipped back to March 11; the new calendar prevents such slippage by omitting leap days in years ending in two zeros, except when the year number is divisible by 400 (e.g., 1600 and 2000). The new calendar was adopted immediately in most Catholic countries, but the Protestant northern parts of Europe did not change over until much later, and in different years in different countries. England held to the Julian calendar until 1753, and by then the difference between dates in Old and New Style, as they were called, had grown to 11 days. Thus George Washington was actually born on February 11, 1732, Old Style, but his birthday is now celebrated on February 22, the equivalent date in New Style.

Kepler has his year begin at noon on January 1. It may sound trivial that January 1 is the first day of a new year, but we find, in fact, a bewildering array of conventions of day and month of New Year's Day, sometimes even different for different purposes at one place and one time (we still have our fiscal years, academic years, and tax years beginning on different dates). March 1 was widely used for that purpose, and in such a calendar September, October, November, and December are, indeed, the seventh, eighth, ninth, and tenth months as their names indicate, and the leap days at the end of February are added at the very end of the year. In England the new year began on March 25 (Lady Day, Annunciation Day) until 1751.

When Kepler begins his astronomical "day" at noon rather than at midnight, he follows a tradition that goes back to Ptolemy. The advantage of this convention is that the observations made during a given night all carry the same date, and it was followed in most astronomical works until the end of 1924, from which time astronomers had agreed to switch to the civil calendar's midnight epoch.

Kepler reckons hours for the meridian of the small island Hveen where Tycho Brahe had observed (ca. $12\frac{3}{4}°$ east of Greenwich), so his noon means noon at Uraniburg, Brahe's observatory on Hveen. The Rudolphine Tables contain a list of terrestrial longitudes—counted from this meridian—(and latitudes) of a great number of important cities, as well as a map of the world with coordinates drawn in, which enables you to convert your local time to Hveen time.

The two parameters given in the mean motion tables for each planet are the heliocentric longitude of the aphelion, $l(A)$, and the mean heliocentric longitude of the planet, $\bar{l}(P)$, which in our notation is

$$\bar{l}(P) = l(A) + M,$$

that is, the longitude of the aphelion increased by the mean anomaly. [Note that $\bar{l}(P)$ would be the heliocentric longitude of the planet if the eccentricity e were zero so the orbit would be circular and the planet's motion uniform with $\theta = E = M$.]

Kepler does not include tables for the earth among his planetary tables; instead we find tables enabling us to find the geocentric longitude of the sun, $\lambda(S)$. In the mean motion tables we find

$$\lambda(Ag) = l(A) + 180°$$

and

$$\bar\lambda(S) = \bar l(E) + 180° = l(A) + M + 180° = \lambda(Ag) + M,$$

where l as before stands for heliocentric longitude, λ for geocentric longitude, S and E for sun and earth, A for the earth's aphelion, and Ag for the sun's apogee, the point farthest from the earth on the sun's orbit.

For each planet, and for the sun, Kepler gives the values of these two parameters for the beginning of every century for an enormous time span. Further, he lists how much these parameters grow in any (whole) number of years up to 100, in the intervals from the beginning of the year to the end of each of the 12 months, and in any number of days up to 31. When year and date are given, by simple addition we can then find $\bar l(P)$ $l(A)$, $\bar\lambda(S)$, and $\lambda(Ag)$.

If the planetary orbit's major axis were truly fixed in relation to the fixed stars, then $l(A)$ would grow precisely at the rate of the precession (Kepler's value is very nearly $51''$ per year). However, he assigns to each aphelion a very slow proper motion so that $l(A)$ grows at slightly different rates for different planets.

For a planet we now find

$$M = \bar l(P) - l(A),$$

and we enter the anomaly tables with this value of M to find θ and r as described above. We immediately get

$$l(P) = l(A) + \theta,$$

but we save the value of r for later use.

For the sun we find likewise

$$M = \bar\lambda(S) - \lambda(Ag)$$

and enter the solar anomaly tables with M to find θ and r. These tables are identical to the ones we would get for the earth if it were treated like any other planet. Again we have

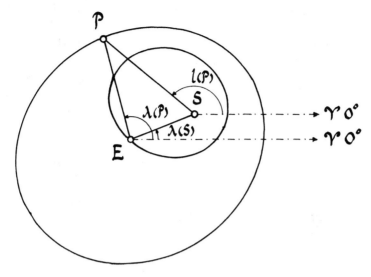

FIGURE 12.

$$\lambda(S) = \lambda(Ag) + \theta,$$

and r is reserved for use in our next and final step, that of finding the geocentric longitude of the planet.

Figure 12 represents the earth E and an outer planet P at a particular moment in their orbits surrounding the sun S. Our aim is to find $\lambda(P)$, the planet's geocentric longitude. We have just seen how we may find $l(P)$ and $\lambda(S)$ as well as the distances SP and ES (they are the values of r). We now consider triangle ESP. We know two of its sides and can find angle PSE, for (see Figure 12)

$$\angle PSE = l(E) - l(P) = 180° + \lambda(S) - l(P)$$

[due to space constraints, $l(E)$ is not marked on the figure]. Thus, we know an angle and the two enclosing sides of the triangle, and we can find its remaining pieces by the standard methods of trigonometry. We particularly want angle SEP, for then we have the desired longitude

$$\lambda(P) = \lambda(S) + \angle SEP,$$

and, at long last we are finished.

So you see that in the end we still have to solve the same old triangle earth–sun–planet that one has had to deal with since at least the time of Ptolemy.

Kepler Motion Viewed from the Empty Focus

We shall now address the problem of how a planet appears to move when observed from its orbit's empty focus—the one not occupied by the sun. Our aim will be, first, to show that the planet moves very nearly uniformly around this focus and, second, to obtain an estimate of its motion's deviation from uniformity. There are, as we shall see, good historical reasons for doing these things.

First let us see, however, in a purely pragmatical way if it is at all worthwhile to tackle the problem. In Figure 13 and Table 4 I compared three models for the planet Mars, whose eccentricity e I have taken to be 0.0933, and whose mean distance from the sun I have used as a unit of length.

Model I is the Kepler ellipse in which the planet P moves about the sun S according to the area law. In Table 4 the columns headed I list the values of r and θ (in degrees and decimal fractions thereof) corresponding to the values of the mean anomaly M given in the leftmost column and produced by Model I. I found the values of r and θ by solving Kepler's equation and otherwise following the procedures set forth above.

In Model II I retained the correct orbit, that is, an ellipse of semi-major axis 1 and eccentricity $e = 0.0933$ with the sun in one focus, but here I let the planet move uniformly about the other focus F with the correct period. I did the latter by letting the angle at F equal the mean anomaly M. In the columns headed II I entered the values of r and θ corresponding to the values of M in the leftmost column, and in the columns I − II are the differences between the correct values of r and θ and those derived

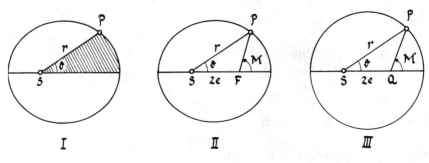

FIGURE 13.

TABLE 4

| M | \multicolumn{5}{c|}{r} | \multicolumn{5}{c|}{ϑ} |

M	r					ϑ				
	I	II	I-II	III	I-III	I	II	I-II	III	I-III
0°	1.0933	1.0933	0	1.0933	0	0°	0°	0°	0°	0°
15	1.0906	1.0907	-0.0001	1.0909	-0.0003	12.51°	12.46°	0.05°	12.47°	0.04°
30	1.0827	1.0828	-0.0001	1.0837	-0.0010	25.15	25.06	0.09	25.06	0.09
45	1.0699	1.0701	-0.0002	1.0720	-0.0021	38.03	37.92	0.11	37.84	0.17
60	1.0527	1.0529	-0.0002	1.0558	-0.0031	51.29	51.17	0.12	51.20	0.09
75	1.0319	1.0321	-0.0002	1.0359	-0.0040	65.03	64.94	0.09	64.98	0.05
90	1.0087	1.0087	0.	1.0130	-0.0047	79.37	79.34	0.03	79.38	-0.01
105	0.9842	0.9842	0	0.9884	-0.0042	94.41	94.45	-0.04	94.49	-0.08
120	0.9603	0.9602	0.0001	0.9637	-0.0034	110.22	110.31	-0.09	110.35	-0.13
135	0.9388	0.9387	0.0001	0.9411	-0.0023	126.79	126.92	-0.13	126.94	-0.15
150	0.9217	0.9216	0.0001	0.9228	-0.0011	144.07	144.19	-0.12	144.20	-0.13
165	0.9106	0.9105	0.0001	0.9109	-0.0003	161.89	161.96	-0.07	161.96	-0.07
180	0.9067	0.9067	0	0.9067	0	180	180	0	180	0

from Model II. How the latter is done I leave as an exercise for the reader. (Hint: Set $FP = \rho$ and

$$\rho = \frac{1-e^2}{1+e\cos M};$$

thus, two sides and an angle are known in triangle SEP, and r and θ can be found.)

In Model III I replaced the elliptical orbit by its circumscribed circle (of radius 1). I placed the sun S at the distance e from the center and made the planet P move uniformly about the "equant point" Q, which lies symmetrically to S with respect to the center, so $SQ = 2e$. The angle at Q is then the mean anomaly M. The values of r and θ derived from this model and corresponding to the values of M in the leftmost column are listed in the two columns III, and the differences between the correct values and these are in the columns I − III. Once again, I leave their derivation as an exercise for the reader. (Hint: Introduce the circle's center C; then $CP = 1$. Triangle CQP then has three known pieces, so the rest of them can be found. Finally, solve triangle SCP.)

The numbers in the difference columns speak for themselves: They are very small, indeed, even the ones in columns I − III. Thus, Models II and III are good approximations to the Kepler-

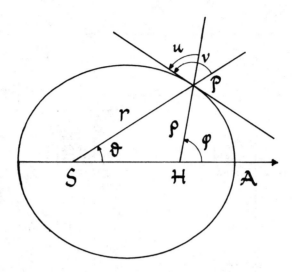

FIGURE 14.

ian motion of Model I, and the excellence of Model III is perhaps particularly surprising, for here we changed not only the manner of motion, but the orbit itself. Here we must remember, however, that an ellipse with small eccentricity deviates only very little from its circumscribed circle. With Mars's eccentricity of 0.0933 and semi-major axis 1, we find for the semiminor axis

$$\sqrt{1-e^2} = 0.9956$$

(for $e = 1/3$, we get 0.943, and 0.866 for $e = 1/2$).

Thus, in this case at least, it seems close to the truth that the planet moves uniformly about the empty focus, and we shall now investigate just how close (or far) it remains so in general.

Let the ellipse in Figure 14 be the orbit of the planet P; the sun S is in one focus, H is the empty focus, and A the aphelion. Let the eccentricity be e and set, as above, the semimajor axis equal to 1. We have then the polar equations of the orbit

$$r = \frac{1-e^2}{1-e\cos\theta} \quad \text{and} \quad \rho = \frac{1-e^2}{1+e\cos\phi} \tag{i}$$

where $r + \rho = 2$, so

$$dr = -d\rho. \tag{ii}$$

If u and v are the angles from the rays from H and S, respectively, to the forward half of the tangent at P (see Figure 14), we have (see page 140)

$$\tan v = r\frac{d\theta}{dr} \quad \text{and} \quad \tan u = \rho\frac{d\phi}{d\rho}. \tag{iii}$$

Because the ellipse reflects a ray from one focus through the other (see page 139), we have that $u + v = \pi$ (we measure in radians, since we use calculus), so

$$\tan u - \tan v \tag{iv}$$

and hence

$$r\frac{d\theta}{dr} - \rho\frac{d\phi}{d\rho},$$

which, with (ii), yields

$$r\, d\theta = \rho\, d\phi.^* \tag{v}$$

The planet obeys Kepler's law, and the area that grows uniformly with time is

$$\text{area } ASP = \mu \cdot T = \int_0^\theta \frac{1}{2} r^2 \, d\theta, \tag{vi}$$

where T is the time elapsed since the planet last passed the aphelion A. We wish to establish a direct link between T and ϕ, and to that end we transform the integral in (vi) using $r = 2 - \rho$ and relation (v):

$$\text{area } ASP = \int_0^\phi \frac{1}{2}(2 - \rho) \cdot \rho \, d\phi. \tag{vii}$$

We now express ρ in terms of ϕ according to (i) and obtain

$$\text{area } ASP = \frac{1}{2}(1 - e^2) \cdot \int_0^\phi \frac{2 + 2e\cos\phi - 1 + e^2}{(1 + e\cos\phi)^2} \, d\phi$$

$$= \frac{1}{2}(1 - e^2) \cdot \int_0^\phi \frac{1 + 2e\cos\phi + e^2}{(1 + e\cos\phi)^2} \, d\phi. \tag{viii}$$

---

*I am indebted to Professor D. T. Whiteside of Cambridge University for having shown me this Newtonian shortcut. Newton had results about the near-uniform growth of ϕ much like the one below. (See his *Principia*, Book I, Prop. XXXI, Scholium, and D. T. Whiteside (ed.), *The Mathematical Papers of Isaac Newton*, Vol. VI, Cambridge, 1974, p. 172.)

Finally, we introduce the mean anomaly M, recalling that

$$\text{area } ASP = \frac{1}{2}\sqrt{1-e^2}\cdot M$$

(when $M = 2\pi$, area $ASP = \pi ab = \pi\sqrt{1-e^2}$), which, with (viii), gives

$$M = \sqrt{1-e^2}\int_0^\phi \frac{1+2e\cos\phi+e^2}{(1+e\cos\phi)^2}\,d\phi. \qquad \text{(ix)}$$

This integral is not at all pleasant, but if we develop the coefficient and the integrand in infinite series according to increasing powers of e, and then integrate term by term, we get

$$M = \phi - \frac{1}{4}e^2\sin 2\phi - \frac{2}{3}e^3\sin^3\phi + \cdots, \qquad \text{(x)}$$

which, when inverted in the rather brutal manner of the 17th and 18th centuries, yields

$$\phi = M + \frac{1}{4}e^2\sin 2M + \frac{2}{3}e^3\sin^3 M + \cdots, \qquad \text{(xi)}$$

so ϕ is equal to M except for terms that contain e raised to the second and higher powers. Thus, the planet moves practically uniformly about the empty focus when the eccentricity is small.

In sum, if we replace

(i) a planet's elliptical orbit by its circumscribed circle, and
(ii) its Keplerian motion by uniform rotation about the empty focus, we shall have achieved an excellent approximation to its behavior, for in either case we have committed errors that involve only second and higher powers of the eccentricity e.

In case (i) we recall that an ellipse's minor semi-axis is

$$b = a\sqrt{1-e^2} \approx a\left(1 - \frac{e^2}{2}\right),$$

where a is its major semi-axis, and in case (ii) we invoke equation (xi). In other words, the kind of motion that Ptolemy employed on the deferent of his equant model is a very good approximation to a Keplerian motion of a planet around the sun with the sun at the observer's place O, the eccentricity of the circle the same as the ellipse's, and the empty focus as the equant point Q.

So we have seen that Ptolemy was on the right track when he abandoned uniform circular motion, philosophically correct though it then was and long remained, and replaced it with a circular motion that is uniform about a point other than the center, but we have not yet justified his planetary models as a whole.

We saw earlier that for an inner planet, the deferent represented the sun's orbit around the earth (the equivalent of the earth's orbit around the sun), while the epicycle corresponded to the planet's orbit around the sun. For an outer planet it was the other way around: The deferent was the planet's orbit around the sun, while the epicycle was the sun's orbit around the earth. With the advantage of hindsight we can see that Ptolemy's planetary models should have been composed of *two* equant models, approximating the model composed of two Keplerian ellipses, for determining a planet's geocentric position shown in Figure 12.

Since that is so, one may well ask how Ptolemy was led to propose a deferent that acts like a very good approximation to a Keplerian ellipse, but an epicycle with a purely concentric motion that is nearly uniform (it is precisely uniform relative to a point on it opposite the equant point) and not like the Keplerian motion.

Before trying to hint at an answer, I should say that the question is partly wrong. Indeed, Ptolemy cannot make an equant model work for Mercury but has to introduce a quite complex crank mechanism to carry the epicycle around. I have avoided mentioning Ptolemy's model for Mercury so far—and I shall not discuss it in detail now—for its complexity places it outside the scope of this little book. The trouble is caused by Mercury's great eccentricity (slightly over 1/5).

The case of Venus is readily disposed of. Its orbit has so small an eccentricity (less than 1/100) that it is already very nearly circular and its motion very nearly uniform. An equant model with the sun's geocentric orbit as a deferent, and an epicycle with Venus's mean distance from the Sun (in a.u.) as a radius, is, then, a very good approximation to the behavior of Venus as seen from the earth.

While Venus seems to be made to fit an equant model, the case of the outer planets is not so simple. I cannot here justify Ptolemy's models for them in detail or try to assess how accurately they yield a planetary position. I shall rather suggest a point of view from which his models look quite plausible.

166   4. Kepler Motion Viewed from Either Focus

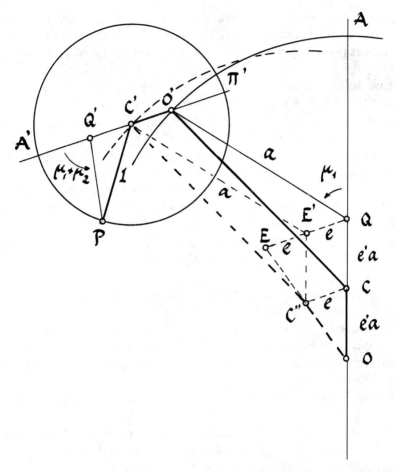

FIGURE 15.

To that end we consider, once again, the model representing the geocentric motion of an outer planet correctly: A deferent that is the equivalent of the planet's orbit around the sun, and an epicycle that corresponds to the sun's apparent orbit around the earth. When we approximate these two components with eccentric circles with equant points, we get an arrangement like the one illustrated in Figure 15.

In Figure 15, O is the observer on the earth, C the center of the deferent whose radius $CO' = a$ astronomical units is the major semi-axis of the planet's elliptical orbit of eccentricity e', and the distance OC is $e'a$ a.u. The point O' moves on the deferent uniformly, with angular velocity μ_1, about Q, which lies on the extension of the line OC, with $OC = CQ$. However, O' is no longer the epicycle's center C', for we now have an eccentric epicycle representing the sun's elliptical orbit around the earth. The apsidal line $\Pi'A'$, on which we have the equidistant points $O'C'Q'$, is of fixed direction, so it moves parallel to itself. The radius of the "epicycle" is 1 a.u., and we have

$$O'C' = C'Q' = e \text{ a.u.},$$

where e is the eccentricity of the sun's (earth's) elliptical orbit. The planet P moves on the "epicycle" uniformly, relative to $C'A'$, around Q'. We note that $\angle QO'A'$ must grow at the uniform rate μ_1 so that the apsidal line $\Pi'A'$ always will have the same direction, and the angular velocity of P relative to $C'A'$ is not μ_2 of Ptolemy's old model, but $\mu_1 + \mu_2$.

It should further be noted that we have scaled the model correctly, in astronomical units, so that it yields not only a fine approximation to the direction from the observer to the planet, but also to its distance OP.

We shall now see that by a slight rearrangement of parts in Figure 15, the planet can be recognized as moving on an epicycle whose center travels on an eccentric deferent. To that end we complete the parallelogram $CO'C'$ and call the fourth vertex C''. Two of the parallelogram's sides are of (constant) length a, the other two are of length e, and these last two sides, CC'' and $O'C'$, have *fixed* direction. The point C'' is then a fixed point, and it can plainly be considered the center of a circle of radius $C''C' = a$ on which the center C' of the epicycle travels.

So an outer planet's geocentric behavior can be very well approximated by an epicyclic model with an eccentric deferent. The eccentricity of the deferent is the vector sum of the eccentricities of the earth's and the planet's orbits (the eccentricities are here considered not as ratios, but as distances measured in a.u.).

This is as close as we can get to a Ptolemaic model for an outer planet with this sort of approximation. Two things are missing for perfect agreement. First, we ought to have found that the epicycle's center C' moves uniformly around the ideal Ptolemaic equant point E that lies symmetrically with O with respect to the eccentric circle's center C'' (see Figure 15). In fact, C' moves uniformly about the midpoint E' of the line joining Q, the old deferent's equant point, and the ideal equant point E, as can readily be seen. Note that

4. Kepler Motion Viewed from Either Focus

TABLE 5

	e	a in a.u.	e'	ae' in a.u.
Mars	0.017	1.52	0.093	0.14
Jupiter	0.017	5.20	0.048	0.25
Saturn	0.017	9.54	0.056	0.53

e: earth's eccentricity;
e': planet's eccentricity.

$$QE'' = E'E = e,$$

the earth's eccentricity.

Second, the planet does not move uniformly with angular velocity $\mu_1 + \mu_2$ relative to a fixed direction, but about the point Q' that is removed from it by the distance e a.u., the earth's eccentricity.

So we have found that a planetary model composed of two equant models, approximating the correct Keplerian ellipses, and Ptolemy's planetary model agree in all but one respect: The centers of uniform motion do not coincide but are displaced—deferent's equant from deferent's equant, and "epicycle's" equant from epicycle's center, respectively—and by the amount e a.u. in both cases, where e as before is the earth's eccentricity. How much that is compared to other dimensions of correctly scaled models is shown in Table 5.

In the *Almagest* Ptolemy deals with the planets in the order we recognize from his cosmology: Mercury, Venus, Mars, Jupiter, Saturn, and the equant model first appears in the *Almagest*'s Book X, where he is concerned with Venus. He first finds its epicycle's radius to be $43;10^p$, if the deferent's radius is 60^p, and that the deferent's eccentricity is $1;15^p = 1/48$, all this by considering Venus's greatest elongation from the sun (we recall that the direction from us to the sun is, roughly, the direction to the epicycle's center); the argument is hinted at in Figure 15. He then determines that the point around which the epicycle's center moves uniformly—the equant point—is a further $1;15^p$ removed from us beyond the deferent's center, so our distance to the equant point is $2;30^p = 1/24$. We have then an equant model where the deferent's center bisects the line from observer to equant point.

Ptolemy now takes for granted that this sort of equant model works for an outer planet as well, with the obvious modification that now it is the direction from the epicycle's center to the planet that is the same as the direction from us to the sun, and he devotes each of the next three books of the *Almagest* to finding the parameters of such a model that best suit one of the outer planets.

The *Almagest* is surely not an autobiographical work, and we find many instances of Ptolemy's changing the sequence of events leading to a certain result for pedagogical reasons or reasons of clarity. Much can be said for Mars, and not Venus, being the planet for which he first discovered the equant model, for (always excepting Mercury) Mars is the planet that, because of its great eccentricity, most clearly shows deviations from mean behavior—after all, it was by an analysis of Mars that Kepler found his laws. Yet it is quite possible that here the *Almagest*'s order of presentation with Venus's as the prototypical model is also the historically correct order of discovery, for, as we saw, Venus seems ideally suited to suggest a deferent with an equant point.

Venus's deferent should be the equivalent of the sun's orbit around the earth, so we should consider Ptolemy's solar model. He takes it over unchanged from Hipparchus, and it is the one example in the *Almagest* of the simplest efficient approximation to Keplerian motion: the pure eccenter, a uniform circular motion viewed from a point other than the center. It works about as well as an equant model if its eccentricity is chosen to be twice that of the Keplerian ellipse—this may be intuitively clear from the above discussions—but only as far as *directions* are concerned. For *distances* it does not work well at all: Thus, in the apsides it would place the sun at the distances $1 \pm 2e$, respectively, where e is the sun's (earth's) eccentricity, and they ought to be $1 \pm e$.

Indeed, Ptolemy's (and Hipparchus's) value for the solar eccentricity is 1/24, and for the eccentricity of Venus's deferent he has 1/48, precisely half, as one would expect if the solar model was to yield correct distances as it must if it is to serve as a deferent in a model that accounts for the variation in Venus's greatest elongation.

What Ptolemy thought about the similarity of his solar model to Venus's deferent is a very good question (which means that I cannot answer it). However, in the *Almagest* they are different in this respect: The sun's apogee is assumed to be tropically fixed, that is, to have constant longitude, while the deferent's is sidereally fixed so that its longitude grows at the rate of the precession of the equinoxes. Incidentally, the assumption that the solar apogee has constant longitude is one of the flaws in the *Almagest* that early Islamic astronomers corrected, as I mentioned earlier.

Selected Bibliography

ACT: O. Neugebauer, *Astronomical Cuneiform Texts*. London 1956. Reissued by Springer-Verlag. The standard edition of all cuneiform texts dealing with mathematical astronomy known at the date of publication.

HAMA: O. Neugebauer, *A History of Ancient Mathematical Astronomy*. 3 vols. Springer-Verlag, 1975. The standard reference work.

Asger Aaboe, *Episodes from the Early History of Mathematics* (New Mathematical Library, Vol. 13). New York, 1963. Kept in print by The Mathematical Association of America.

Asger Aaboe, "Observation and Theory in Babylonian Astronomy." *Centaurus* (1980), Vol. 24, pp. 14–35.

Bernard R. Goldstein, "The Arabic Version of Ptolemy's Planetary Hypotheses." *Trans. Am. Philos. Soc.* N.S. 57, 4 (1967).

Hermann Hunger and David Pingree, *Astral Sciences in Mesopotamia* Brill, Leiden, 1999. An expert and detailed survey of Mesopotamian astrology and astronomy with a full bibliography.

Alexander Jones, "*Astronomical Papyri from Oxyrrhynchus.*" *Mem. of Am. Philos. Soc.* (1999), Vol. 233. An epoch-making publication that changed our ideas about Hellenistic astronomy.

E. S. Kennedy, Colleagues and Former Students, *Studies in the Islamic Exact Sciences*. Beirut, 1983. The startling first papers on the Marāgha School and its followers are among the included essays.

O. Neugebauer, *The Exact Sciences in Antiquity*. 2nd ed., Providence, 1957. (Dover, 1969). The classical monograph on ancient mathematics and astronomy.

Olaf Pedersen, *A Survey of the Almagest*. Odense Univ. Press, 1974. A very useful, clear, and detailed summary and analysis.

N. M. Swerdlow and O. Neugebauer, *Mathematical Astronomy in Copernicus's* De revolutionibus. 2 vols. Springer-Verlag, 1984. A detailed account of Copernicus's complicated astronomical models.

G. J. Toomer, *Ptolemy's Almagest*. Springer-Verlag, 1984. The definitive English version with commentary.

Christopher Walker (ed.), *Astronomy before the Telescope*. British Museum Press, London, 1996. An excellent collection of essays written by experts for a general audience.